Construction Claims

A Short Guide for Contractors

Paul Netscher

"I've been involved with over 120 projects and all have had Variation Claims – some for millions of dollars, almost doubling the project value. Variations are an inevitable process with construction projects – there will be changes and delays no matter how well they are planned and executed. Yet, more than 99% of my Claims were settled amicably and most of what we claimed was approved. More importantly we completed further projects with these same employers." Paul Netscher

Copyright Note
Copyright © 2016 Paul Netscher
All rights reserved. No part of this publication may be reproduced or transmitted, in whole or in part, by any means without written permission from the publisher.
Published by Panet Publications
PO Box 2119, Subiaco, 6904, Australia

www.pn-projectmanagement.com

ISBN: 978-1537086323

Available from Amazon.com and other retail outlets

Legal Notices

It should be noted that construction projects are varied, use different Contracts, abide by different restrictions, regulations, codes and laws, which vary between countries, states, districts and cities. Furthermore various industries have their own distinct guidelines, acts and specific protocols which the contractor must comply with. To complicate matters further these laws, acts and restrictions are continually evolving and changing. Even terminologies vary between counties, industries and Contracts and may not be the same as those included in this publication. It's therefore important that readers use the information in this publication, taking cognisance of the particular rules that apply to their project.

Each project has its own sets of challenges and no one book can cover all the steps and processes in every project. Every Contract Document is different (even minor word and punctuation differences can have profound impacts) so it's important to understand your Contract. This publication covers a broad range of Claims and some might not be relevant for your project. Some of the author's personal opinions may not be pertinent to certain projects, employers or companies. Readers should undertake further research and reading on the topics particularly relevant to them, even requesting expert advice when required.

Therefore, the author, publisher and distributor assume no responsibility or liability for any loss or damage, of any kind, arising from the purchaser or reader using the information or advice contained herein.

The examples used in the book should not be seen as a criticism of people or companies, but, should rather be viewed as cases which we can all learn from. After all we've all made mistakes. Any perceived slights are unintentional.

Cover layout by Clark Kenyon, www.camppope.com
Cover photograph by Paul Netscher

Preface

In the course of most construction projects delays and variations are encountered which give rise to Variation Claims. These Claims can be between the employer and the contractor, or between the contractor and their subcontractors. Construction Claims are a vast and complicated subject with differences which depend on the Contract Document and the law of the country where the Contract is administered. Often the contractor isn't granted all the time, or money, that they are entitled to, and frequently these Claims end in lengthy and expensive disputes where both parties end as losers.

It's important that contractors present their Variation Claims timeously and in a fashion that's hard to refute. In this book I've looked at Variation Claims from the contractor's view point, looking at reasons for lodging a Variation Claim, gathering the supporting documentation, what to include in the Claim, negotiating the Claim, and finally avoiding Claims. No single book can cover every form of Contract Document, or even every event leading to a Claim. Then, pursuing a Claim through the courts and legal system can sometimes yield different results between apparently similar Claims, which may be dictated by legal principals rather than understanding the events on the construction project.

Therefore this book serves only as a guide, a starting point to developing your Claim. I've tried to present the information in an easy to read format, staying away from legal terminologies which can sometimes be confusing. Your particular Contract Document will form the basis of your Variation Claim, which will be supported by the supporting documentation, specifications and drawings and the approved Construction Schedule, together with project correspondence, Instructions and meeting minutes. Often it pays to engage professionals to provide advice on the Claim – especially on large and complicated Claims.

Variation Claims shouldn't be seen as a way for the contractor to make up for under-pricing the project, or to make up their losses due to poor project management. Spurious Variation Claims cost time and money and divert energies from where they could be better used on the project. False Claims are seldom successful, often only irritating employers and possibly alienating them, meaning they won't use the contractor on future projects.

Unfortunately some employers have a view that all Claims are spurious and the contractor is simply out to make money. This view often 'blinkers' the employer and they fail to see how their actions disrupt the project, often leading to animosity and disputes. Resolving Claims with these employers can often be a lengthy, time consuming and costly process for all parties, with few winners except their legal teams.

Acknowledgements

Thank you to the many readers who've purchased my previous two books. Your reviews have inspired me to write this book which I hope will be equally useful. Thank you also to the readers of my articles on LinkedIn and my website. Some of the articles you found helpful and commented on have been incorporated into this book.

Cliff thanks for suggesting I write a short book. Sorry this isn't quite what you had in mind.

Thank you to the many people I've worked with in the construction industry and those who have helped me amicably solve hundreds of Claims.

Thanks Sandra for your support.

Contents

Introduction xi
Variation Claims xi
Why contractors don't always claim their additional costs xii
Reasons Claims are rejected xiii
How this book will help you xiv
Notes for reading this book xv

Chapter 1 – Types of Claims 1
Extension of Time Claims 2
Concurrency and Precedence of delays 8
Force Majeure 10
Reasons for Extension of Time 11
What is access? 17
What information and by when? 18
Disruption Claims 20
Standing Time Claims 21
Acceleration Claims 23
Constructive Acceleration 24
Variation Claims 25
Negative Variations 27
Nil Impact Variations 28
Rate Variations 28
Conflicting clauses 29
Design and Construct projects 30

Claims against subcontractors 31

Insurance Claims 32

Claims for non-payment of invoices 32

Cardinal Change 33

Contract Termination 34

Suspension of the Contract 35

Wrongful Termination 36

Use of the Omission clause 36

Time at Large 36

Liquidated Damages 37

Claims by subcontractors against contractors 38

Claims by the employer against the contractor 38

Summary 40

Chapter 2 – Supporting Documentation 42

The Contract Document 42

Tender drawings (Drawings issued with the request for price) 44

The Construction Schedule (Programme) 44

Schedule Float 46

Resource driven Schedules 47

Updating the Construction Schedule 47

Documentation 48

Instructions 49

Weather records 50

Project meeting minutes 51

Daily Logs or Diaries 51

Personal Diaries 52

Request for Information (RFI) or Engineering Queries 52

Emails 54

Letters 54

Photographs 55

Drawing Register 56

Distribution of drawings 56

Project Reports 57

Searching for evidence of claimable events 57

Is the contractor's team really on their side? 59

Records of Variation work 60

Hand-over documentation 60

Claim mitigation 61

Subcontractors 61

Insurance 63

Archiving records 64

Summary 65

Chapter 3 – Preparing the Claim 66

The Variation Claim process 66

Variation Claim Notification 68

Time-Bar 68

Put the effort in 69

What should be included in the Claim 70

Don't assume an earlier Claim will be approved 71

Extension of Time Claims 72

Factors to consider when formulating an Extension of Time Claim 74

What is a reasonable duration for additional tasks? 76

The spatial arrangement of structures and delay impacts 78

Changes and their impacts on access 78

Deciding on which Claim to focus on 79

Acceleration Claims 80

Costs for Extension of Time Claims (prolongation costs) 81

Mitigating losses 84

Pricing Extension of Time overheads using the Contract Preliminaries ... 85

Costs to include in Variation Claims 86

Disruption costs 89

Formulating disruption costs using the 'measured mile' 90

Overhead and profit on Variation costs 91

Inflating Claims 92

Cost calculations 92

Have the full impacts of the Variation been considered? 93

Pricing the Variation using existing prices 93

Pricing the Variation using Cost-plus methods 94

Validity of the price 96

Supporting Documentation 96

Checking the Claim submission 97

Why is the Claim so expensive? 98

Termination 98

Insurance Claims 99

Claims from subcontractors 100

Summary 101

Chapter 4 – Negotiating and Winning Claims 103

Why disputes can be harmful 103

How to avoid disputes 104

The cup of coffee approach 105

Employers that don't respond to Variation Claims 105

The impartiality of the employer's Agent 107

Claim Registers 108

Variation Claim meetings 108

Negotiating a Claim 110

Bribery and corruption 111

Variation Orders or Contract Amendment Orders 112

When disputes are unavoidable 112

Understand the laws in the state or country 114

Revising the Construction Schedule 114

Summary 115

Chapter 5 – Avoiding Claims and Disputes 116

Why Variation Claims aren't always good for contractors 116

Ask these important questions 118

Is there an alternative? 119

Avoid certain employers 119

Avoid particular Contract Documents 120

Understand the tender or pricing documentation 121

Visit the project site before submitting the price (tender) 121

The Construction Schedule 122

Team work 123

Communication 124

Understanding the employer 125

Verbal communication 125

The contractor's team – the contractor's early warning 126

Training 126

Subcontractor pricing 127

Selecting the subcontractor 128

Subcontractor documentation 129

Managing subcontractors.. 130

Managing Design and Construct Projects....................................... 131

Shop drawings 132

Samples and mock-ups 133

The win-win approach 133

Take it higher if necessary 134

Bogus Claims.. 135

The role of the contractor's management 136

Ask questions... 137

Take reasonable steps to prevent Claims from arising 138

Give notice before the Contract Value is exceeded 139

Don't exceed the value of Variation Orders................................. 139

International Contracts 139

Ask for help 140

Summary... 140

Conclusion 142

Glossary.. 145

Notes.. 150

References.. 151

Introduction

These are newspaper headlines from a recent construction project in Australia
- **Samsung C&T reveals $1b loss on Roy Hill** (Sydney Morning Herald)
- **NRW counts losses at Roy Hill** (The West Australian)
- **Roy Hill contractors' dispute resolved, Samsung to pay $30m to NRW holdings** (ABC News)
- **The Supreme Court has allowed Roy Hill project manager Samsung C&T to seize a $7.5 million security bond from construction company Laing O'Rourke** (The West Australian)
- **A judge has allowed Roy Hill iron ore project manager Samsung C&T to seize $68 million in security bonds from Spanish manufacturer Duro Felguera** (The West Australian)
- **Roy Hill Holdings has escalated its row with Samsung C&T by attempting to seize a $235 million performance bond from the project manager** (The West Australian)
- **$1b at stake in Roy Hill mine row** (The West Australian)

From these headlines it is clear that some construction projects can become messy with huge Claims and counter Claims. Companies can lose millions of dollars and inevitably the only winners are lawyers.

Variation Claims

Have you had a Variation Claim rejected even though you believed it was totally legitimate? You probably blamed the employer for being unfair? But is this the full story?

Construction projects invariably have some changes and Variations. These may be due to the employer changing the Scope, altering the project specifications, delaying the contractor in some way or not fulfilling their contractual obligations. Sometimes the project site conditions aren't as per the Contract Document. These changes invariably increase the construction company's costs so it's vital to submit a Claim to recover them.

In addition some Variations may delay construction work, resulting in the project being completed late. Not only will the contractor incur additional costs to remain on the project longer but the employer may impose Damages on the contractor for late completion. In addition projects that are completed late often reflect poorly on the contractor's reputation.

I've been involved with over 120 construction projects and only a few haven't increased in value – with some almost doubling in value. Some Variations have been worth millions of dollars. Many of my projects also suffered Delays and it has been necessary to claim an Extension of Time. Fortunately, with the exception of only two projects, these Variations were resolved and approved amicably. In most cases we received close to the full value claimed, or were granted the entire Extension of Time requested. Even the two cases that weren't resolved amicably were resolved in a relatively short time, without going through the full Dispute Resolution processes available in the Contract Documents. Furthermore, in almost every case we remained on good terms with our employer and went on to construct several more projects for them.

Unfortunately many contractors aren't as fortunate and we sometimes hear accounts of project disputes that take years to resolve, going through lengthy legal and Dispute Resolution processes, with, in some cases, neither party gaining a satisfactory result. These disputes cost time and money to argue. They often end in acrimony between the parties which negatively impacts the project as well as future relationships. They are also often in the media, tarnishing the reputations of those involved.

Why contractors don't always claim their additional costs

I was astounded a few years ago when I joined a small construction company to find they weren't claiming Variations and were carrying out work that wasn't in their original scope without claiming the costs, or time, they were entitled to claim. They simply weren't aware they were entitled to lodge a Variation Claim for this additional work.

Other contractors don't claim for Variations because they fear they'll damage the relationship with their employer. Sometimes they believe their employer will look after them at the end of the project so it's not necessary to submit Claims. Unfortunately employers don't always look after their contractors, even if the contractor has been 'Mr Nice Guy'. In fact, in most instances the employer doesn't even know that the contractor is being 'nice' and hasn't claimed a Variation that they were entitled to claim.

It sometimes gets worse when the project is finished late (because of the employer), since if the contractor hasn't claimed and been granted an Extension of Time they are penalised by the employer and forced to pay Damages for late completion. Unfortunately when projects end badly most in the team are quick to apportion blame, protecting their own jobs, and forgetting who played 'Mr Nice Guy'.

Some other reasons construction companies don't submit Variation and Delay Claims are:

1. The contractor's Project Manager isn't familiar with the project Scope of Works and what was included in the contractor's tender or bid price.
2. The Project Manager hasn't read the Contract Document.
3. The Project Manager isn't taking note of the work their team is doing, or when and what the employer has provided.
4. The contractor's person preparing the invoices to the employer isn't familiar with the construction site conditions, or the construction work that's been completed, so doesn't invoice for additional items of work.
5. The contractor doesn't have the required knowledge or training to understand that they are entitled to be paid a Variation.
6. The construction Project Manager is focussed on delivering the project and neglects the financial aspects.
7. The construction Project Manager believes financially everything will work out fine at the end of the project so it's not necessary to submit Claims.

It's essential the contractor's team understands the Contract Document and Scope of Work. When there are changes and delays these should be claimed in terms of the Contract Document.

Contractors cannot afford to construct work for free, or be penalised due to their employers not fulfilling their obligations or causing delays to the project.

Reasons Claims are rejected

Unfortunately even when contractors submit Variation or Delay Claims they are sometimes rejected. This might be because:

1. The Claims aren't legitimate for reasons including:
    a. The Variation is due to the contractor's fault.
    b. The contractor should have allowed for the event in their price since it was a known fact when they priced the project, or it was the contractor's risk.

2. The Claims are submitted late, or aren't in accordance with the Contract Document.
3. The Claim is poorly put together without supporting evidence.
4. The employer doesn't understand the Claim.
5. The employer automatically assumes the contractor is out to 'steal' from them and assumes the Claim is bogus or inflated.
6. The employer doesn't believe they are wrong.
7. The employer expects the contractor to be submissive and carry out work for free.
8. The Claim is based on the wrong information or it has factual, Schedule or arithmetic errors.

Unfortunately rejected Claims are a waste of effort – effort that could have been focussed on managing the project. It's therefore imperative that contractors only submit Variation Claims when they have a legitimate cause, and not simply to try and unfairly gouge money from their employer, or to recover money they've lost on the project due to under-pricing or because of poor management on their part.

When preparing the Variation Claim the onus remains on the contractor to present the facts of their Claim in a logical and clear manner, with all supporting evidence, in such a way as to convince the employer that the Variation has genuine merits.

How this book will help you

In this book I'll outline some reasons for Claims, what we need to consider when formulating our Claims and what we should include in our Claims. I briefly discuss resolving the Claim with the employer and how and why we should avoid Variation Claims where possible. Exactly how you word your Claim and what's included may vary depending on the Contract Document.

It should be noted that I haven't referred to a particular Contract Document as there are many forms. Furthermore Employers often add their own special conditions to every project. I've therefore taken a very general approach. This approach could change depending on your specific Contract Document. Even a minor alteration of a single clause can dramatically alter the contractor's Claim. It's therefore best to read and understand the specific Contract conditions applicable to the project. What worked on a previous project may not work on the next. What was disallowed on one project may be acceptable on the current project. Contractors shouldn't assume that they, or their team, will automatically understand the Contract Document because it's the same (or similar) Document used previously. If in doubt ask an expert for advice.

This book serves only as a guide to contractors for preparing their Claims.

Notes for reading this book

In this book I refer to the contractor and the employer as usually being the two parties in the Claims process. The employer may be the client, owner, customer, developer, home owner or the Agent employed by the owner to manage the construction project. The Agent could be the client's Project Manager, the client's Engineer or their Architect. The contractor is the company appointed to construct the whole facility or part of the facility. The contractor may employ subcontractors to construct parts of the work that are in the contractor's scope.

Often the contractor has cause to lodge a Claim against the employer. However in some cases the contractor may also submit a Claim against their subcontractor. If an insurable event occurs the employer, or contractor, or subcontractor may be able to lodge an insurance Claim.

Subcontractors sometimes have reason to claim from the general contractor or main contractor employing them. In this case, readers who are subcontractors would in this book become the contractor and the general contractor would be referred to as the employer.

Terminologies vary between countries, even between companies and amongst different construction fields (building, civil, electrical and mechanical). Please refer to the Glossary at the end of the book for an explanation of the terms I've used in this book. But more importantly, refer to the particular project's Contract Document to see the terminologies used, and ensure these are understood.

I've tried to keep this book short and simple, steering away from legal terminologies. Contractors unfortunately usually don't read lengthy legal tomes. I want this to be a quick reference guide that can easily be referenced for those small 'every-day' Claims. For more complex and larger Claims it's usually advisable to seek advice and help from experts.

Chapter 1 – Types of Claims

Variation Claims arise for a number of reasons. There may be delays caused by either the employer or their team, or due to outside factors. Employers may issue construction drawings which differ in some respect from the drawings priced by the contractor. Often employers issue Instructions or Variation Orders which instruct the contractor to carry out additional work, change what they have already completed, or that changes information contained on a construction drawing.

Contractors need to be aware when there is a change to the information they originally priced and the information which is detailed in the Contract Document. It's vital to read through and understand the construction Contract and the Bid Documents (Tender or Request for Price documentation), as well as the related correspondence. It's important that the contractor regularly compares construction drawings and specifications with those issued at bid stage and included in the Contract Document to ensure that they haven't changed.

Contractors should continually ask the questions:
1. 'Are we constructing what we bid or quoted for?'
2. 'Are the site conditions as expected at bid or pricing stage?'
3. 'Has the employer timeously fulfilled all their obligations in the Contract Document?'
4. 'Are we being delayed by an event in our control or as a result of our actions? Could this event have been foreseen when we priced the project and would it have been reasonable to allow for it in our Schedule?'

If the answer to any of these questions is no there may be reasons to lodge a Variation Claim.

The main types of Variation Claims are:
1. Extension of Time Claims.
2. Acceleration Claims.
3. Increase in costs.
4. Damages

However contractors also need to be aware that subcontractors, suppliers and the employer can also lodge Claims against the contractor. They need to manage the project to minimise the chance of these Claims against them arising, and ensure they have the documentation and knowledge to defend these Claims when necessary.

Extension of Time Claims

Construction projects can be delayed for a number of reasons. Sometimes the contractor causes the delay themselves – obviously this isn't reason for a Claim against the employer.

Whether the contractor can lodge a Claim for a particular delay event will depend on the following:

1. Is the delay due to a cause that wasn't included in the Contract Document? Often Contract Documents outline specific delay events which the contractor should allow for. These events could include:
    a. Restricted project working hours.
    b. Interfacing with the employer's activities.
    c. Complying with the employer's security arrangements to access the project site.

   Obviously if the Contract Document clearly articulated that the contractor must make provision for these events then the contractor can't argue later that these events delayed them unless the actual events were worse than originally set out in the Contract Document or explained to the contractor.

2. Is the delay on the Critical Path? The Critical Path is made up of the items on the Construction Schedule which form the shortest path from the start of the project to the end of the project. A delay to a task on the critical path will delay the completion of the project.

    For example: the construction of a house may consist of a number of activities including foundations, walls, roof, and the finishes. Clearly if the roof is delayed the house cannot be completed. The same house construction may include a perimeter wall around the property. The contractor may elect to construct this wall at the start of the project to provide security for the project. But clearly the house can be built separately while the property perimeter wall is being constructed. In terms of the Schedule the perimeter wall can start at any time providing it's completed before the house is completed. So if the house's Construction Schedule was 40 weeks long, and the perimeter wall takes 2 weeks to construct, the wall could in theory start in week 38 without delaying the project. So even if the Construction Schedule showed the wall starting in week 1 and the information to enable the construction of this wall arrived in week 15 and the procurement of materials for the wall took 2 weeks, clearly the perimeter wall would still be completed long before the end of the overall project and there would be no delay to the project. The contractor may be inconvenienced by not being able to start the wall when they wanted. This inconvenience might include providing temporary security fencing around the property and may mean they don't have continuity of resources (such as bricklayers), but they don't have reason for an Extension of Time Claim since the wall is not on the Critical Path. They may however be entitled to submit a Variation Claim for the other reasons outlined above.

However, it should be noted that an item that wasn't on the Critical Path can be delayed to such an extent that it now becomes Critical.

For example: in the previous example if the information required to construct the perimeter wall was received at the end of week 38 and the procurement of the materials took 2 weeks then clearly the perimeter wall could no longer be completed by the end of week 40, which was the original project duration. Allowing for the procurement period of 2 weeks for the materials and another 2 week to construct the wall means that the project will be completed 2 weeks late and the contractor will be entitled to 2 weeks Extension of Time.

3. The delay must impact the task on the Critical Path.

    For example: in the above example the contractor wouldn't need to know the paint colours until a day or two before painting was due to start – which might only be after week 30. The contractor may request the paint colours on week 2 and only receive them on week 29, but work hasn't been delayed. Of course if the house was in the countryside, far from a paint supplier, the contractor may require the paint colours a week before painting is due to start (and not a day before as in the city) to allow for transporting the paint from the city to the project site.

This brings us to another issue which is the Lead Time for materials. Materials usually take time to be ordered and delivered. Specialist windows could, say, take 9 weeks to fabricate. This Lead Time depends on transport, fabrication time, time to prepare the design and manufacturing drawings and usually time for the employer or the contractor to approve the design and drawings. It's important that the employer is made aware of these Lead Times so they can provide information sufficiently ahead-of-time before the scheduled start date for the task, to enable the item to be procured, manufactured and delivered to the project.

Also, it should be noted that sometimes abnormal weather events occur (say heavy rain) which don't actually impact the project (or certainly not the Critical Path activities), because the activities are within a weather proof environment.

> *For example: if we are building a new rail tunnel, heavy rain might not impact the tunnel boring machine unless the tunnel was flooded. Invariably the tunnel boring would be on the Critical Path. The rains may impact the activities outside the tunnel but if the rain delay isn't sufficient to make these activities critical the contractor probably won't have cause to claim a weather delay.*

4. The delay must be an unexpected event.
    a. Most construction projects are negatively impacted by poor weather. It could be cold, storms, rain, wind and even extreme heat. Often these are normal weather patterns and the contractor should have allowed for the impact of this adverse weather in their Construction Schedule. Sometimes we have extreme weather events which are out of the ordinary. Often in these cases the contractor may be able to claim for the weather extremes beyond that which is the norm.

       > *For example: during the month of May the region normally experiences 50mm of rain over 4 days, but during construction the project actually experienced 100mm of rain over 8 days in May. If activities on the Critical Path were delayed by the 8 days of rain the contractor should be able to submit a Claim for 4 of those days, which are the rain days in excess of what would be normally expected during that month.*
       >
       > *However it should be noted that if the project receives rain for 4 days in June but the average conditions for that region was for 6 days of rain in June, then, the employer may argue that the 4 days the contractor claimed in May could be offset by the 2 days of better conditions in June.*
       >
       > *It is important to note though that the employer cannot shorten the overall project duration because the contractor experienced more favourable weather conditions than could have been*

expected. So in the above example, if in June the project received no rain the employer may be entitled to offset the 6 days of better than expected weather in June against the 4 days of worse weather in May and reject the contractor's Claim for a 4 day Extension of Time. But, they cannot lodge a Claim against the contractor for a 2 day reduction in time, being the 6 days of additional good weather in June less the 4 days of worse weather in May.

It should be noted that these impacts are usually only restricted to the direct time lost because of the extraordinary weather. Damages to the project as a result of the weather are usually claimed against insurance. (See later.)

b. The contractor can't claim for known statutory holidays, but they would have cause to submit a Claim if the country declared and legislated an additional holiday or non-working day.

Case study: I worked in a country that held their government elections on a Wednesday every 5 years and always declared the day of the election a statutory public holiday. On our projects we claimed an Extension of Time for this additional holiday. Some employers argued we should have allowed for the holiday since we knew when we priced the project that there would be an election that year. We argued successfully that the precise date of the election wasn't known and it was possible that the government could declare the election on any day over a six month period and that it was possible for the election to have fallen outside the project period.

5. The delay event cannot be related to complying with current legislation or existing good practices known at the time of pricing the project. For instance:
   a. The contractor can't claim a delay for carrying out normal quality inspections and tests.
   b. They also can't claim a delay for complying with safety legislation, or good safety practices.
   c. They shouldn't claim a delay for clearing equipment and materials through customs unless there is an additional unexpected disruption beyond their control which delays this process.
6. The delay must be due to something outside the contractor's control and something they couldn't have expected or allowed for. So for instance if the contractor is responsible for the design they cannot claim against

their employer if the design drawings are late unless the employer has unnecessarily delayed approving the design or has requested design modifications which would make the design different to what was contained in the Contract Document. Also, if the contractor damaged a known water pipe, or if a water pipe the contractor installed ruptured, and the water from the pipe flooded the works causing a delay, then the contractor wouldn't be able to claim for the delay because the flooding was attributable to the contractor's actions. Obviously if the contractor couldn't have foreseen that there was a water pipe in that location and took due precaution then they could not be held responsible for damages to the pipe and they could be able to submit an Extension of Time Claim.

7. The delay event must be an event that hasn't been specifically excluded in the Contract Document. Unfortunately some Contract Documents are very one-sided and employers push risks onto their contractors. Employers may push the risk of extreme weather events, or unforeseen ground conditions, onto the contractor by inserting specific clauses, or references, making the contractor responsible for the delays and costs which resulted from these events. Most Contract Documents exclude events such as war, natural disasters and terrorism.

8. The employer must be aware of the obligation the contractor claims they haven't fulfilled. For instance, as discussed above there are varying Lead Times for materials and equipment. The employer needs to be aware of how far in advance information is required for tasks on the Construction Schedule.

    For example: if, in the previous example of the house, the window information is required 9 weeks ahead of when the windows are scheduled to be installed the contractor should have disclosed this Lead Time in their bid submission (quotation), or at the very least early in the construction process, and certainly long before the information is required. If they haven't they can't lodge an Extension of Time Claim when they receive the window drawings, say 5 weeks before they are due to be installed..

9. The delay can't be due to work stoppages or strike action by the contractor's workers or one of their subcontractors unless this was part of a national work stoppage which the contractor couldn't have mitigated or avoided.

10. The delay can't be because of shortages of material or equipment unless these are caused by national, or significant regional, events such as work stoppages, or flooding. However, if the items are specialist and are only

available from a few suppliers and the contractor has taken all mitigating actions, including placing orders at the earliest possible opportunity there could be cause for claiming a delay if the items cannot be supplied for reasons beyond the contractor's control.

11. The delay must impact the approved Construction Schedule. If the contractor is ahead of the approved Construction Schedule they cannot claim a delay if the employer didn't provide access and information in accordance with their actual progress, but, which was supplied ahead of the dates on the approved Schedule.

    For example: if the contractor is 10 days ahead of Schedule they cannot expect the employer to provide drawings and access 10 days earlier than the approved Schedule. In this case if the information was issued 12 days after the contractor needed it for them to remain 10 days ahead of Schedule the actual Extension of Time the contractor could claim is only the 2 days the information was late according to the approved Construction Schedule.

12. There being no other delays which have already occurred which impacted the task. (See Concurrency of delays below).

13. The delay must be for something the contractor could not have reasonably foreseen. Now this can be a 'grey' area and some employers like to refute Claims using the following words; 'an experienced and knowledgeable contractor should have foreseen or expected to encounter the obstacle or event'. Some employers seem to expect their contractors to be almost clairvoyant and yet, they think they can act blindly to the possibility of any Variations occurring and don't have to undertake their own research when preparing the Contract Document. However, the contractor does take responsibility to acquaint themselves with the physical conditions on the project site and neighbouring properties that may hinder construction, as well as considering the local conditions – which include the expected weather conditions, local taxes, availability of skills and materials and so on.

    For example: if rock is obviously present on the project site, or in neighbouring land, then encountering rock during construction wouldn't be an excuse for an Extension of Time. Again it isn't always clear how much research the contractor should have performed during the tender stage. Certainly they couldn't have been expected to carry out a detailed geological survey of the project site. So rock which is encountered during construction that impacts the contractor's progress and which wasn't obvious on the project site during pricing and which wasn't highlighted in the employer's pricing documentation would be reason for the contractor to lodge a Claim for Extension of Time as well as for additional costs encountered for excavating the rock.

14. Usually the delay can't be a consequential delay. In other words if a storm strikes the project the contractor may be able to claim an Extension of Time for the time they couldn't work during the storm. However, if the storm destroyed completed work – say high winds ripped off the roof, then it's unlikely the contractor can claim for the time to replace the roof and other damages the storm inflicted on the building.

Of course I'm not advocating contractors don't lodge a Claim for some of these events. Certainly each Contract is different and Claims which some Contract Documents disallow are acceptable in terms of other Contract Documents. Contractors should seek expert advice when they are uncertain. Sometimes employers aren't as contractually savvy as they should be, so on occasion it may still pay to submit a Claim. However, be aware that you may be wasting both your and your employer's time and it may impact your integrity when you later submit Variation Claims you are entitled to.

Concurrency and Precedence of delays

Some delays and events occur concurrently, and sometimes these are a result of faults by both the employer and the contractor, and occasionally even a Force Majeure event (see later). It's often difficult to resolve who is responsible for the overall delay and to quantify it. It may, therefore, be necessary to put the individual delays together to see their overall result. Some Contract Documents provide a ruling in the event there are concurrent delays caused by both the contractor and the employer, but unfortunately most Documents are silent leaving the parties, and in many cases the courts to decide how concurrent delays should be treated. Unfortunately in many cases there's no clear precedent how concurrent delays should be treated.

When considering the cause of the delay, the norm is to attribute the cause to whoever caused the first delay, with the second delay only coming into effect once the first delay is over.

For example, if the contractor delayed themselves by 3 days due to material arriving late, and on the second day the employer delayed the contractor by 10 days because a drawing was issued late, the delay the contractor could claim would be 10 days minus 2 days, (the contractor had caused themselves the first delay of 3 days and 1 of these had elapsed when the employer caused a delay, in other words the delay due to lack of materials overlapped the employer's delay by 2 days), resulting in only an 8 day delay due to the employer's actions.

However, if the employer first caused the delay of 10 days due to late information, and 3 days after the employer's delay the contractor delayed themselves by 3 days due to their materials arriving late, then the delay caused by the employer would be 10 days. (The delay the contractor caused themselves occurred during the period the project was already delayed by their employer's late information.) Of course most contractors will in both cases always try and claim the full 10 days delay caused by the employer.

Obviously this would vary from project to project, and in many instances some common sense, and give-and-take, should prevail from all parties.

It should also be noted that in certain cases courts have ruled that the contractor is entitled to the full duration of the delay caused by events outside the control of the contractor even if they overlapped with preceding delays caused by the contractor.

In other cases the dominant cause of the delay has been taken as the delay that's used to decide the amount of the Extension of Time. If the dominant cause is due to the contractor's delay there won't be an Extension of Time granted for the portion of the employer's delay that overlapped with the contractor's delay. However, deciding the dominant cause is often problematic.

But, on other projects if the project is delayed by concurrent delays caused by both the employer and the contractor the contractor is granted an Extension of Time for the full period the employer delayed the project, but the time granted for the period which overlapped the delay caused by the contractor is without payment for their costs.

It's important for contractors to note that in the event the employer has caused a delay event that they don't relax, slow down progress or divert resources so they fall behind Schedule. Slippage caused by the contractor could then be interpreted as a concurrent delay.

For example: the employer provides information late for a follow-on task so the contractor knowingly slows their progress on the preceding tasks in the knowledge that they only have to be completed in time for when the employer provides the missing information. The contractor essentially falls behind on the Construction Schedule. The employer could later argue that the late information wasn't the critical event that delayed the project, but rather it was the contractor who completed the preceding tasks late. Should the contractor's Extension of Time Claim go to arbitration or litigation a third party would be presented with the contractor's evidence of the late information from the employer, and the employer's evidence that the contractor hadn't completed the preceding tasks so couldn't have used the information if it had been issued on time. In this case the contractor faces the risk of their Extension of Time Claim being rejected, or possibly being viewed as concurrent and then only being granted the Extension of Time without compensation for their costs.

Should the contractor decide to slow down a task, for instance to provide continuity of their resources for when the employer did provide information, then it would be prudent to write a letter to the employer giving them notice they were deliberately slowing progress and the reasons why they were doing so.

Force Majeure

This is a term given to delays which none of the contracting parties' control, or is responsible for, such as unusual adverse weather conditions and national worker strikes or work stoppages. Normally the Contract Document makes reference to Force Majeure, and the contractor may be entitled to an Extension of Time covering the delays incurred due to the event. However, in most cases each party carries their own costs which resulted from the delay.

The Contractor must be aware of which events are Force Majeure, and which are not, since it's often more beneficial to submit an Extension of Time Claim for another reason and under another clause.

For example: on one of our projects the employer's personnel participated in a work stoppage across all of their facilities which prevented us from working on the project for a number of days. Some contractors on the project claimed it was Force Majeure, we maintained that as it was only the employer's personnel (even though it was national across all of their facilities) it was a work stoppage which the employer could have prevented (even though their workers' demands may have been unreasonable). We claimed that the employer failed to provide access to the work area which prevented us from working, and claimed an Extension of Time for the delay with costs.

As with most delays the contractor will have to demonstrate that they attempted to put mitigating actions in place.

For example: if there's prior warning of a national transport strike the contractor must be able to demonstrate that they made every attempt to stock-pile materials and get critical items to site ahead of the strike.

Reasons for Extension of Time

There are a number of reasons for claiming an Extension of Time providing they meet the criteria previously outlined. They could include:

1. The employer issues an Instruction to stop work for reasons unrelated to the contractor. The Instruction may be due to the employer rethinking their design, or to allow the employer's work, or the work of their contractors, to proceed. In some cases employers may issue an Instruction to the contractor to stop work because of a fault by the contractor such as; unsafe working conditions, poor quality work, or failure to obtain design approvals or the required construction permits. In these cases the contractor can't apply for an Extension of Time unless they can prove that the employer was wrong to stop the work and that the contractor was in fact compliant.

2. The employer provides construction information late. I'm sure we've all been on projects where the employer's drawings have been issued late. This delays the project and frustrates the contractor's team. If there's an approved Construction Schedule the employer should know when information is required and there's no excuse for it being issued late. Providing an Information Schedule (a schedule of when the employer must provide information to the contractor), which links back to the Construction Schedule and which allows for procurement Lead Times is a valuable aid.

3. Late access. Often contractors are dependent on the employer providing access to the work area in accordance with the agreed Construction Schedule. The project will be delayed if these dates aren't met, and the contractor is entitled to claim for these delays. Care should be taken when accepting access to a work area that the area is safe (unless it has been specifically stated that the contractor must make the area safe). Also the area given to the contractor should meet the specified dimensions and heights as laid out in the Contract Document.

    For example: if the employer is handing over an excavation (which has been previously excavated by the employer or a contractor engaged by the employer) to the contractor it should be excavated to the specified dimensions and depths, and the sides of the excavation must be safe with no threat to the contractor's workers from ground and rocks falling in.

    If the excavation doesn't meet these requirements the employer should be notified so that they can rectify the deficiencies. Failure to do so may mean that the contractor does additional work at an extra cost and which invariably requires extra time.

It's vital that when access is granted to the contractor that they check the access is compliant and immediately reports problems to the employer.

For example: if the contractor receives access several days before it's required, but they fail to check the access granted is acceptable until the day work is scheduled to begin. If the contractor only then finds fault with the access, and claims for the delay while the employer rectifies the problems, they may find the employer has reason to reject the Extension of Time Claim as they will assert that if the contractor had notified them of the problem immediately they were granted access it could have been rectified immediately and the contractor wouldn't have been delayed.

4. Changes in specification. This is something that contractors don't always detect until it's too late. Change in specifications often increases the price of items, but the items with the new specifications could also have longer manufacturing Lead Times. In some cases specifications are changed after items have been ordered which means that the original order has to be cancelled and new orders placed which can significantly delay projects. Sometimes customers and their designers aren't aware of the implications of these changes to the project. If they are immediately made aware they could revert back to the original specifications to avoid delays and additional costs.

5. Scope increases.

Case study: we constructed a large dam which was priced on a re-measurable bill of quantities. During construction our construction team was struggling to keep to the Construction Schedule and continually asked for more resources. Eventually we had much more equipment and people than we had allowed when we priced the project so consequently were showing large losses on our cost reports. When the team eventually caught up with re-measuring the quantities they found that some items had increased in quantity by more than 25%. Unbeknown to the construction team they had been accelerating the works at our cost – doing more work in the same period of time.

Many projects increase in Scope and contractors need to continually compare the actual Scope with the Scope they priced. Increased Scope usually means we need more time to complete the project, or additional resources to complete it in the original time frame. Contractors need to timeously notify their customer of Scope increases as these usually add to the project cost which is detrimental to the customer's budget as well as possibly impacting the project's Completion Date.

Some Contract Documents specify that a Contract must increase in Scope by a defined percentage (often 25%) before an Extension of Time Claim can be lodged. But in any case the contractor couldn't claim for an Extension of Time for minor increases in Scope. They also couldn't claim for Extension of Time for an increase in Scope which doesn't impact the Critical Path – they could claim for the costs associated with carrying out the additional work but not for additional time.

6. The employer's activities cause delays. Sometimes the Contract Document includes customer activities and constraints that the contractor has to work around and accept. In these cases the Construction Schedule should take these impacts into account. However, often during the course of construction the employer introduces new constraints.

> *For example: On some mining projects we had to shut-down the work several times a week to allow our employer to carry out blasting activities. In other cases our work hours have been limited. Some existing facilities had particularly rigorous security arrangements (not covered in the Contract Document) which delayed the movement of people and materials onto the project. These interruptions impacted our work, causing us delays which we successfully claimed Extension of Time for.*

7. Additional quality tests and inspections. The employer's quality managers, at times, can introduce additional tests or quality inspections which weren't mentioned in the Tender Documents or the Contract Document. These can add additional costs and cause delays. Some employers add in additional 'hold' or inspection points, or require 24 hour, or even 48 hour, notice periods for inspection. It's important the employer is immediately notified of the impact of these requirements so they can reassess these requirements to ensure the additional costs and time is worth the benefit.

8. Late drawing or design approval. The Construction Schedule and Contract Document should stipulate the maximum turn-around time for the employer to approve the contractor's drawings and designs. Some employers aren't good at keeping to these times. However, in some cases contractors cause further delays which can't be claimed because their

drawings aren't correct, are not submitted through the correct channels or aren't in the correct format.

9. The employer's team doesn't immediately respond to Requests for Information (RFI's) and drawing queries. Unfortunately I'm sure we've all received drawings with missing or conflicting information. On occasion querying and receiving the corrected information can be a tedious and time consuming process which delays the project. These delays need to be brought to the employer's attention. We always include a list of all outstanding queries in our project meetings.

10. Employers' not providing facilities and utilities in the required quantities and in the time they were obligated to supply them.

    Case study: on one of our earthmoving projects the employer undertook in the Contract Document to supply water for construction at a specified point and in the quantities we required. Unfortunately the designated water point wasn't ready until we were several months into the project. The alternative supply was 5 kilometres further away, it was used by other contractors and it couldn't deliver the quantity of water we required every day. We not only needed more water tankers to haul the water a longer distance, but they had to queue behind other contractors to refill. As we know water is required for ground compaction so the limited water supply reduced the amount of material we could compact each day. This reduced our production, which increased our costs and caused major delays to the project.

    Ensure that the employer provides the utilities and facilities where the Contract Document says they should, and that the quantities supplied are in accordance with the agreed Contract.

11. The employer's other contractors impact and delay the contractor's work. They may restrict access to the contractor's work areas, damage completed work or hold-up the contractor's work where they are required to interface with them. Even work outside the immediate work area could dramatically impact the contractor's work.

    For example: the employer's contractors may excavate a trench across a road which could cut-off access to the work area, or cause delays to the contractor's equipment that can't move freely past the obstruction, or the contractor may have to take a longer route to reach their work area to avoid the area that's been closed. All of which can cause delays and increase the contractor's costs

12. The employer revises drawings which require the contractor to redo work which is already completed, or the contractor has to re-order materials or equipment, thus delaying the project while the items are procured.
13. The employer changes the sequencing in the Schedule. The employer might decide they want some sections of the project earlier and others later. This at first sight might not increase the overall duration of the project, however this changed sequencing may impact access to work areas, or it increases congestion in the work area, which causes a delay.
14. Unforeseen site conditions. These could include:
    a. Existing services (such as; water pipes, electrical and data cables, and gas mains) which weren't known and couldn't have been foreseen when the project was priced. These may require further detection and uncovering work as well as their relocation which can delay the scheduled work.
    b. Difficult ground conditions which weren't known and couldn't have been foreseen at bid stage. This could include:
        i. Different soil conditions which may require a different foundation design to what was allowed for in the contractor's price and Construction Schedule.
        ii. Rock which might be more difficult and slower to excavate.
        iii. Unsuitable materials (such as clay) which have to be removed and replaced with better material.
        iv. Contaminated ground containing materials such as asbestos or dangerous chemicals which have to be removed and disposed of by specialists.
        v. A high water table on the project site which impacts excavations which may be inundated by this water.
        vi. Discovering an archaeological artefact on the project site which has to be protected, excavated and removed by a specialist.
        vii. Uncovering of buried building rubble which has to be removed.
        viii. Discovering previously unknown cavities – sink-holes or old mine workings. These may have to be filled or

          structures might have to be relocated to avoid the cavities.
   ix. Ground characteristics which are different from those expected by the employer. For instance some ground or rock could be more unstable than expected which could result in deep excavations having to be made wider with flatter slopes to safely accommodate the characteristics of the ground.

   It's pertinent to note that if the contractor is responsible for the design, as well as undertaking a detailed geological survey, then sometimes they might not be entitled to an Extension of Time for some of these events, although it would depend on the terms of the Contract Document.

15. Extreme weather events, such as one in fifty year floods and catastrophic storms. The claimable weather conditions could be occasions when the weather is more severe than the average climate for the region. More rainfall, more snow, more windy days, more violent winds, or more extreme temperatures during construction than the average (for the same months) for the area where the project is. Some Contract Documents may allow the contractor to claim for all weather related delays so it's important to understand the Contract fully. However, in most cases the contractor can only claim for the direct time lost due to weather above the recorded averages.

16. Items that the contractor specifically excluded in their bid submission and which were included as exclusions in the Contract Document.

    For example: when we operated in an area prone to cyclones we specifically excluded cyclones from our price. We didn't know whether we should allow for 1 cyclone or 6 or more in a season. Each cyclone would disrupt the project for at least 3 days. We could have allowed for say 6 cyclones in our price, but if there was only 1, we would have profited and the employer would have paid for cyclones that didn't occur. By specifically excluding cyclones and making them a claimable event our employer only recompensed us for the actual time lost when cyclones impacted the project.

    Of course it doesn't help to just specify these events in your bid documentation, but they must also be included in the final agreed Contract Document – something that many contractors forget to check.

17. Change in legislation. This could include the country declaring a special holiday. Changed legislation could also result in a change in the building materials which might not, for instance, comply with new fire legislation.
18. National strikes, work stoppages or a disruption which causes the contractor's employees to stop working, or impacts the supply of critical materials, such as fuel. This could also include disruption by the employer's employees which interrupts progress on the construction project. It should be noted if the contractor chose to purchase an item from another country, disruptions in that country which delays shipping the item probably wouldn't be considered a legitimate claim for Extension of Time unless the employer had specified the item had to be purchased from that supplier, or when there was no alternative supplier.
19. Employer supplied equipment or materials arriving late, being damaged when they arrive on the project, or not being fit for purpose. However, it should be noted that the onus is normally on the contractor to check that the employer supplied item conforms to the specifications and dimensions. Contractors would be expected to check the item when it arrives on site and not wait until they are about to install it.
20. Errors on employer issued drawings which either result in work being redone, or causes delays while clashes, missing information and problems are resolved.
21. Drawing coordination problems. Frequently Architect's and Engineer's drawings have conflicting information, or the mechanical, electrical and plumbing drawings have clashing services or utilities. Sometimes the Engineers or Architects haven't allowed for the utility services, so holes have to be cored through concrete slabs or broken through walls. All of this rework and these clashes create delays and additional costs.
22. The employer failing to obtain statutory permits which prevents the contractor from carrying out work on the Critical Path.

What is access?

Employers sometimes grant access to a portion of works, but there's no access to that work area – the area is available but the contractor is unable to move people and equipment safely to the area to enable them to execute the work. Access should mean that the area:
1. Is safe to work in, unless the Contract Document specifically states that the contractor must first make the area safe.

2. Meets the contract specifications – it's the correct dimensions, at the correct height (within the specified tolerances) and the quality meets specifications (including passing the required tests).
3. Can be accessed safely unless the Contract Document requires the contractor to build the access.
4. Is clear of any obstructions unless it's a requirement for the contractor to remove these obstructions.
5. Is legally the employer's, unless, the Contract Document makes provision for the contractor to acquire these permissions. Contractors cannot start work on land which doesn't belong to the employer unless there are written permissions in place from the legal owner allowing work to proceed.

> *For Example: I have been handed access to work areas where the employer or their contractors were still working, or they had materials stored there which blocked our access. In other cases we have been given access to areas that the employer should have given us graded to a specific level and compacted to a specified limit, which they weren't. In these cases we refused to take access.*
>
> *Sometimes employers only hand over a portion of the area they should give the contractor. They then expect the contractor to start work on part of the section and then claim they haven't delayed the contractor. In some instances this may be acceptable to the contractor and they may be able to maintain working in accordance with the Construction Schedule. Unfortunately this isn't always the case and the contractor often experiences disruption and frustrations to their work.*

It's important that there is a written record of when access is granted which is signed by the parties. Claims can be won or refuted with this document.

What information and by when?

There's information we can use and then there's information that's really no help at all. I'm sure we've all had construction drawings issued to us that have areas clouded with notations that the clouded section is on 'Hold'. The employer's team will maintain that the drawing has been issued on time, but the fact that the contractor couldn't use the information because it was marked 'Hold' sometimes seems to elude them.

The employer's favourite is to claim the contractor could have started work on the sections that weren't on Hold. This work may only take a couple of days and completing it could disrupt the work plan the contractor was going to follow, often creating more of a hindrance than helping the contractor remain on Schedule.

Contractors must insist employers provide sufficient information in the required time and in accordance with the Construction Schedule to start the work utilising their resources efficiently. This information should be for areas which won't later hinder construction of the sections on Hold.

Unfortunately, sometimes drawings issued to contractors are of a poor quality with numerous errors or missing dimensions. I'm sure we have all, at some stage, been in the field, ready to start, only to open the drawing and find we have insufficient information to start. A dimension is missing, or there are errors with the setting out information provided.

> *For example: we had a surveyor come specifically to the project to set out a structure only to find that the information provided would have placed the structure 2 kilometres away from its intended position. We had to send the surveyor away and they had to return at a later date, after the employer supplied the correct information. Of course there was a further delay as the surveyor couldn't return immediately we received the correct information due to their other work commitments.*

Employers will always argue that the contractor should have checked the drawings before they were ready to start. But, is it really the contractor's job to check that the employer's drawings are correct? I would argue no. Nonetheless, I always endeavour to check construction drawings as soon as they are issued to ensure there is sufficient information to start construction.

> *Case study: on one project, in frustration after receiving numerous drawings with errors I returned an error laden drawing to the employer with a letter saying I refused to accept the drawing since it had so many errors. The quality of drawings improved after that.*

Furthermore, the information must be received in sufficient time to enable the contractor to plan the work, procure the equipment and materials, and get them to the project site. As previously discussed the contractor should have specified Lead Times for information required from the employer when they priced the project, and these should have been included in the Contract Document.

> *For example: even if the employer provided the construction drawing 3 weeks ahead of when a section of work is scheduled to begin, but neglected until 2 weeks later to provide the missing dimensions to the steel bolts shown on the drawing which had to be incorporated into the section of works, and if the bolts required a 3 week procurement time the work*

could be delayed. However, the bolts might not be required immediately as other work included on the drawing may have to be completed first. The delay in providing the missing information would apply directly to the Schedule task relating to installing the bolts. In this case it might have been 1 week after work on the section started – which means the missing information was only received 2 weeks before that and not the 3 weeks that was required – resulting in a 1 week delay.

Disruption Claims

A Disruption Claim is a claim for the interference of the contractor's activities and work, or a loss of productivity, resulting from an action or event caused by the employer or their representatives or due to an event outside the contractor's control which couldn't have been foreseen. Disruption Claims are sometimes included with Delay Claims but they are usually different. Disruption may not result in a delay, particularly if the items impacted are not on the Critical Path.

Disruption (loss of productivity) sometimes results in a delay to the work being carried out and not necessarily a delay to the completion of the project. The work produced is not necessarily changed, it simply takes longer to complete or requires additional or different resources from those the contractor priced and allowed for, which then results in the contractor incurring additional costs.

Possible causes of disruption could include:
1. The employer's operations interfering with the contractor. This could include the contractor having to stop their work to allow the employer's operations to continue in the area.
2. The employer requesting frequent urgent Variations resulting in the contractor having to move their workers from the tasks indicated on the Schedule to the Variation work. Productivity is best when workers can concentrate on one task without interruption.
3. The employer restricting access to the work area. This might include limiting the equipment able to work in the area, longer access routes to work areas or positioning laydown areas further from what was envisaged by the contractor at tender stage.
4. The contractor having to interface with the employer's workers and contractors. This could result in congested work areas, additional dangers caused by the employer's work (such as cranes working over-head or cutting, grinding and welding activities), obstructions to the work area and damage to the contractor's completed work which they have to repair.

5. Multiple errors on the employer's drawings. These result in work crews standing while the errors are resolved and while Supervisors reset-out the work.
6. Most delays caused by the employer could also result in a Disruption Claim as well.

It should be noted that in some cases the employer may have specified in the Contract Document that the contractor should allow for interfacing with their operations, activities and contractors. I've worked in existing factories and facilities where the employer's operations were carrying on all around us. Productivity can be reduced by up to 50% and usually additional supervision is required. However as these conditions were known when we priced the project we had no cause to claim additional monies.

Disruption Claims are often difficult to substantiate unless the approved Construction Schedule is fully resourced and that the contractor has kept accurate daily records of the personnel and equipment actually used on those tasks impacted by the disruption. The contractor also has to demonstrate that they've done everything reasonable to mitigate costs.

Standing Time Claims

Standing Time claims relate to costs the contractor incurs when their employees and equipment on the project aren't able to work due to a cause attributable to the employer. Standing Time is really a Disruption Claim although it is referenced as a specific claimable event in some Contract Documents.

Example 1: if the employer issues an Instruction ordering the contractor to stop work because of safety violations caused by the contractor, then the contractor wouldn't have cause to claim for Standing Time unless the contractor could prove that their work was safe and the employer's Instruction was incorrect.

Example 2: if there is a heavy rain which floods the work areas and the contractor, for instance, cannot work for two days due to the areas being flooded they usually cannot claim Standing Time costs for this time – they may be able to submit an Extension of Time Claim. If the flooding was a result of the employer's actions then the contractor would be able to claim Standing Time as well as an Extension of Time and all the resulting costs associated with the flooding of the work areas.

Example 3: if the employer stops the contractor one hour before the end of shift because the employer is carrying out blasting operations then the contractor may be able to claim the costs of their people and equipment that cannot work because of this Instruction.

Example 4: if the employer doesn't give the contractor access to the project site on the Schedule start date and the contractor isn't able to work in accordance with the Schedule then the contractor may be able to claim the standing costs of equipment and personnel already mobilised to the project which cannot be utilised. But, should the contractor have personnel and equipment ready to mobilise to the project the employer usually won't pay for them because they aren't on the project site yet. Also, employers usually won't pay for the resources if the contractor mobilises them to the project after knowing they don't have access to the work area. The contractor literally has to keep these resources 'on the shelf' at their cost. But, should the contractor be able to argue that they are going to lose the resources and wouldn't be able to find similar replacement resources if they didn't pay them they may have a successful Claim for the costs of these resources – in particular for specialist equipment and subcontractors, or rare skills.

It should however be noted that these costs usually have to be proven costs.

So in example 3 above; if the contractor lets their employees go home an hour early without paying for the hour they didn't work then the contractor couldn't claim the costs of their employees' wages as these costs weren't incurred by the contractor.

Some Contract Documents reference Standing Time and in particular that the contractor cannot claim Standing Time unless the time exceeds a specified minimum time – often two hours.

For example: if the contractor's team couldn't work for 1 hour they couldn't claim any Standing Time if the minimum time specified was 2 hours. But if they were prevented from working for 5 hours they could claim 3 of the 5 hours. (Although this will depend on the exact wording of the clause in the Contract Document, because in some circumstances the contractor may be able to claim the full 5 hours as Standing Time.)

In addition to the Standing Time claim the contractor could claim an Extension of Time if the period they couldn't work impacted the Schedule completion date.

The contractor is normally only entitled to claim Standing Time costs of employees and equipment directly associated with the works. Costs of employees and equipment paid under the overheads would be recoverable in the Extension of Time claim. Having said this though, contractors can normally only claim an Extension of Time on a daily basis, so it's unlikely that if the contractor was forced to stand for one hour that this would lead to a one day's Extension of Time.

However, if the contractor is forced to stop work for one hour every day over the period of a week they may find that they've lost a day's progress on the Schedule and could then submit a claim for a one day Extension of Time.

Standing Time often occurs when a section of work is halted because of lack of access or insufficient information. This is where Standing Time claims become difficult. Employers will usually argue that the resources on the affected portion of works could be used elsewhere. Yes, they can be utilised elsewhere, but would it be productive? Can you cram more resources into the other areas and are the resources right for the work?

If the contractor intends to lodge a Standing Time Claim it is essential the employer is notified immediately the contractor's resources are standing. Accurate records of the time and costs need to be maintained and agreed with the employer daily.

Standing Time costs are those directly related to the equipment and personnel standing whilst they are prevented from working. If an item of equipment is standing the contractor cannot claim for fuel, maintenance or wear and tear. Also the contractor can only claim the minimum hours they have to pay for their equipment and personnel.

> *For example: if the contractor's employees have to be paid a minimum of eight hours in a day the contractor cannot claim ten hours from the employer, even if the project normally works ten hours every day. The employer only has to pay the time the resources couldn't work up to the minimum hours the contractor has to pay for them.*

Acceleration Claims

Acceleration Claims occur when the employer requests the contractor to accelerate the Construction Schedule. The fact that the contractor finishes the project earlier without a formal request from the employer to accelerate the project isn't reason for an Acceleration Claim.

Acceleration Claims are normally a result of:
1. The employer asking the contractor to complete the same amount of work in a shorter time.
2. After the employer grants an Extension of Time because of delays or an increase in Scope, the employer then requests the contractor to complete the project by the original Completion Date, or in a time shorter than the Extension of Time granted. In this case it's important to get the employer to first agree to the Extension of Time and then to the Acceleration Claim.

Acceleration Claims are often difficult to quantify. Firstly though, is for the contractor to determine if it's possible to accelerate the project and complete the work within the time frame requested by the employer. It's unwise for the contractor to agree to accelerate the project if the requested completion date isn't achievable, since the agreed Acceleration Schedule will become the new Construction Schedule and the contractor will be liable for Liquidated Damages (or other Claims from the employer) if they don't complete the project by the agreed date.

In some instances the contractor may be behind Schedule through their own fault. In this case, if the employer instructs the contractor to accelerate to catch up the time lost because of the contractor's fault this won't be cause for an Acceleration Claim and the contractor will be expected to accelerate at their own cost.

Constructive Acceleration

In some cases, despite the contractor requesting an Extension of Time for a justifiable delay the employer either fails to respond to the Claim or rejects it. Often in these cases the employer makes it clear (in meetings and/or correspondence) that they expect the project to be completed by the stipulated Completion Date. Failure to complete the project by this date will result in the imposition of Liquidated Damages against the contractor.

In this scenario the contractor can either ignore the employer, continuing as per the adjusted Construction Schedule (allowing for the delays) in the assumption that ultimately their Extension of Time claim will be granted. In this case they run the risk of having Liquidated Damages deducted from them when they don't achieve the Completion Date and they also won't be able to recover their costs associated with the Extension of Time Claim until the Claim is agreed. Even if the Extension of Time is ultimately granted the dispute could drag on for years (long after the Liquidated Damages have been deducted) disrupting the contractor's cash flow. It should also be noted that where the employer specifically instructs the contractor to work to the existing Construction Schedule despite the Extension of Time Claim, and the contractor doesn't comply with the Instruction it could provide cause for the employer to Terminate the Contract if the contractor failed, without reasonable excuse, to proceed with due expedition to complete the works by the required Completion Date.

The alternative is to accelerate the works and claim the acceleration costs. Of course this presumes that it's possible to accelerate and achieve the Completion Date.

Where an Extension of Time is refused and the contractor believes they are entitled to the extension they should:
1. Provide any substantiation to their Claim that the employer has requested. If the contractor thinks the employer's reason for refusing the extension is flawed they should lodge a Dispute in terms of the Contract.
2. Provide the employer with a revised Construction Schedule accelerated to achieve the Contract Completion Date despite the delays. This should be accompanied with a letter noting that this is an Acceleration Schedule. The letter should also include the Acceleration Costs associated with achieving the Schedule.
3. As an option, until such time as either the Extension of Time is granted or the contractor is instructed to accelerate the works, submit a letter to the employer noting that Time is at Large. However, there are other risks associated with stating Time is at Large (see later), so careful consideration must be given to this step.
4. Keep accurate records of all the resources employed in accelerating the project.
5. Vigorously pursue their Extension of Time Claim since this is the proof that acceleration was required.
6. Claim acceleration costs incurred in the monthly valuations, thus ensuring that the employer is well aware of these costs.

Basically the contractor is placed in the invidious position that they could be penalised if they don't accelerate and also penalised if they accelerate without a formal Instruction. Care must be taken to ensure all documents are lodged correctly and where necessary approach an expert for additional advice.

An alternative approach is to discuss with the employer how the Construction Schedule can be modified in such a way that the employer is given sufficient Partial Access so their follow on work isn't delayed, while the overall Completion Date for the project is extended to accommodate the delay or additional work. This could be a win-win solution for both employer and contractor. Approval of the revised Schedule should be sufficient to claim the costs related to the new Completion date.

Variation Claims

Most construction projects will change and vary from the works that were originally priced. Contractors must ensure they are paid for all the additional work they have to execute. Variations occur for a number of reasons including:
1. Changes in specifications.

2. Additional Scope – extra work which wasn't allowed for in the employer's Tender Document (Request for Price, RFP) and consequently wasn't priced by the contractor. This could include new items as well as a change in the quantities of the items priced in the original Contract.
3. Errors and omissions in the Contract Document – resulting in the contractor not pricing or allowing for certain items, restrictions and specifications which they now have to allow for.
4. Changes to construction drawings – this may include executing additional work, or having to redo work already completed.
5. Delays resulting in an approved Extension of Time Claim.
6. Changes in working conditions such as:
    a. Encountering unexpected ground conditions, for example excavating in rock which wasn't expected and allowed.
    b. Encountering hazardous materials on the project site which could delay the project and require specialists to remove.
    c. The discovery of artefacts on the project site.
    d. The unexpected presence of underground water which could delay the project, require specialist equipment and even design modifications.
    e. Encountering unexpected utility and service lines such as; gas mains, water pipes and electrical cables which have to be rerouted, or accommodated in the building design.
7. Changes of commercial or contractual terms and conditions. These could impact when the contractor is paid, the amount of retention money withheld, warranties required including changes to the warranty period, amongst others.
8. Drawing errors and drawing coordination problems which result in rework or Extension of Time Claims. Sometimes service lines clash requiring pipes that have already been installed to be rerouted to accommodate other services. Often engineering drawings differ from architectural drawings, or are at odds with mechanical and electrical drawings. These clashes and variances inevitably result in delays, rework and additional works. I'm sure we have all received Instructions to core holes through concrete slabs and cut holes through walls to accommodate pipes and ducting which weren't allowed for.
9. Changes of law within the state or country which increases the price of goods such as new taxes and duties on goods and services.
10. The employer or their contractors damaging completed work.
11. Instructions issued by the employer.

Negative Variations

Not all Variations result in an increase in the contract amount. In some instances a Variation may result in a decrease in the project's price. This might be because:
1. The Scope of Work is reduced. It should be noted, as discussed later, that the employer cannot remove Scope and give it to another contractor without mutual agreement. Nevertheless, in certain cases the employer may remove work for budgetary reasons or if planning approvals aren't received.
2. The employer changes the specifications to a lesser specification.
3. The employer specifies a different product or item which is cheaper than the original materials.
4. The employer elects to supply an item that the contractor was supposed to supply. This however shouldn't be because they are able to procure it cheaper elsewhere.

It should be noted that generally employers cannot request a negative variation for events such as:
1. More favourable actual weather conditions occurring compared with those which would normally have been expected.
2. Better ground conditions than was anticipated at tender stage.
3. The contractor completing the project in a shorter time (unless this was due to the employer removing Scope).
4. The contractor elects to use an alternative construction methodology which is cheaper or faster than the one originally envisaged. However, in some cases the employer may agree to modify their design to accommodate these construction methods on condition that the contractor passes some, or all, of the savings to the employer.

Also, when the employer reduces the Scope, or supplies some materials, the contractor may decide not to give back the full value of the work removed. Contractors may be entitled to claim:
1. Loss of profit for the reduced Contract Value. In other words they might not return the value of their profits on the removed items. Although if the overall Contract Value has increased due to other positive Variations they possibly shouldn't be claiming loss of profits for items removed. It's often difficult to substantiate the amount of profit the contractor originally had on the omitted items. Unless there's a specified percentage for profit in the Contract Document the contractor and employer have to agree on a percentage considered the norm for this type of work.

2. Expenses they've already incurred in procuring materials and items which are now omitted. If the materials are already on the project these could include transport and handling.
3. Subcontractor's expenses which may include their loss of profit and their expenses incurred to date.

Mostly contractors don't like processing negative Variations. Who likes giving money back? In general I prefer processing negative Variations (although reluctantly) since it establishes the contractor's credibility. Even when negative Variations aren't disclosed they're often uncovered by the employer before the end of the project. When the employer uncovers it first they may think the contractor has been dishonest in not highlighting the negative Variation. Furthermore, if the employer calculates the value of the negative Variation they'll invariably calculate a value higher than the contractor is willing to give back, and the contractor then has difficulty in reducing the value the employer is requesting back.

Nil impact Variations

Even when a Variation has no impact it's sometimes useful to submit a zero impact Variation. It closes the Variation out, providing the employer with certainty (they're not waiting expectantly for a Variation still to come) and it often demonstrates to the employer that the contractor appears to be fair. Employers like to think they are receiving something for free which may make them more amenable when the contractor later submits a positive Variation.

This doesn't mean that the contractor should necessarily be submitting nil impact Variations for new drawings which don't impact their price, but rather nil impact Variations may apply to Instructions issued by the employer which have to later be reconciled to Variation Orders.

Rate Variations

Sometimes a Variation is as simple as submitting a revised rate for an item or task because the description of the item priced at tender stage has changed, for example because:
1. The height or dimensions have changed.
2. The quantity increases. The original supplier might now be unable to supply all the material, which could mean alternate more expensive suppliers must be used.
3. The specification is different.

4. The item is required in a smaller quantity than was originally priced. Suppliers may have discounted the price when large quantities were required and these discounts no longer apply. Also, the transport costs could be the same for the lesser quantity as it was for the original quantity, meaning that the transport cost per unit now increases.
5. The item is to be installed in a different manner from that allowed, or expected, during the bidding process.

    For example: ceramic floor tiles are changed to be fitted in a diagonal pattern which involves more cutting and is more difficult than tiles fitted in a square pattern.

Rate Variations particularly apply to Re-measurable Contracts. In some cases there isn't a rate for a particular item and the contractor has to submit a new rate.

Conflicting clauses

Unfortunately often Contract Documents have conflicting clauses or information. Even during the execution of the work conflicts can arise between drawings and specifications. Some Contract Documents make express provision as to the order of precedence of the various Documents which makes their interpretation easier.

In the case where the precedence isn't clear the norm is that the order is as follows:
1. Modifications to the Documents.
2. The Agreement.
3. Addenda, with those of a later date having precedence over those with an earlier date.
4. The Supplementary Conditions.
5. The General Conditions of Contract.
6. The main Specifications.
7. Drawings and supplementary Specifications.
8. Other documents specifically itemised in the Agreement as part of the Contract Documents.

Where there are conflicts within a specific document normally the more burdensome clause prevails. If the clauses have the same weight the standard would be that the clause that applies would be the clause most favourable to the party that wasn't responsible for preparing the Contract Document. In other words it would be ruled against the party that prepared the Contract.

It should also be noted however that a clause that is commercially unreasonable, or one that could produce an absurd result, would be ignored or over-ridden by other clauses. (What is commercially unreasonable or absurd may have to be decided by a court.)

Clauses that violate public policy or conflict with the laws of the state or country aren't acceptable.

In some jurisdictions it can be argued that when there are two conflicting clauses in the document then the one that appears first would be the one to be applied – in particular this could be used if none of the above arguments apply.

Design and Construct projects

Sometimes the contractor has to design and construct the project. They are responsible for designing the project in accordance with the brief and Scope included in the Contract Document. The contractor usually appoints a design team which may consist of one or more experts such as Architects and Engineers from various disciplines. In some cases the contractor may have the expertise within their organisation. Unfortunately some contractors don't manage the design process very well. The onus is on the contractor to ensure that:

1. Their design team delivers the design and drawing timeously so that construction isn't delayed and that they allow sufficient time for the employer to check and approve the design and drawings where this a Contract requirement.
2. The design complies with the design brief and that the finished facility will perform in accordance with the employer's requirements and that it will conform to the employer's specifications.
3. The design will ensure that the completed facility will comply with the applicable codes and laws of the country and the region, unless the employer has obtained relaxation of these for the facility.
4. The design takes into account their preferred construction methods.
5. The designers have allowed for the local conditions, the skills and resources available, and specify local materials where available.
6. The design doesn't add additional features, or have a more onerous specification, than the employer requested unless these are necessary to comply with local laws, codes or the project permit conditions.

The contractor may have cause to claim against their design team if the design doesn't meet the design brief or if it's late.

However, the employer can cause additional costs or delay the project during the design process. These could be because:

1. Employers delay approving the design and the construction drawings, taking more time than allowed in the Contract Document and Schedule.

2. During the design review process the employer, and in particular their operations team, add in items they want included in the design which weren't included in the original brief. Not only does this cause delays with the design which has to be modified to incorporate these changes before being reviewed again, but it invariably adds additional costs for the contractor.
3. The employer changes the design brief or specifications.

In addition the project site conditions might not be as expected or described in the Tender Documents which could require the design to be modified. As previously described there could also be unknown existing services or unexpected ground conditions which need to be accommodated, resulting in design modifications and consequent delays and additional costs. It should however be noted that if it is the contractor's and their design team's responsibility to carry out investigations of the project site, including geological investigations, then encountering conditions which entailed the design having to be modified would generally not be reason for a claim against the employer.

It's important the contractor has a suitably qualified and experienced manager to oversee the design process to ensure the design remains on track and within the original project Scope. Both the designers and the employer must be immediately notified when the process risks being delayed, or when additional Scope is added, so that action is taken to minimise the Variation and also so the employer understands that their actions have increased the project cost, or perhaps delayed the overall project.

Claims against subcontractors

Now I'm not suggesting that contractors should be lodging Claims against subcontractors simply to make money from them. On the contrary, contractors should be doing everything to support them. However, from time to time subcontractors don't fulfil their contractual obligations, or the contractor incurs additional costs because of the subcontractor. Some of these costs could include:
1. Services supplied to the subcontractor, which the subcontractor should have allowed for and provided, such as:
    a. Accommodation for the subcontractor's employees.
    b. Transport for the subcontractor's employees.
    c. Clearing the subcontractor's rubbish.
    d. Access scaffolding provided to the subcontractor.
    e. Lifting equipment and offloading facilities provided for the subcontractor's materials and equipment.

  f. Provision of water or power to the subcontractor.
  g. Providing offices and other site facilities.
  h. Plant and equipment.
  i. Personnel to assist the subcontractor.

  It should be noted that in some cases the Subcontract Document may make provision for the main contractor to supply some, or all of the above, so it's essential the Subcontract Documents are read and understood.

2. Repairing work completed by the contractor or other subcontractors which has been damaged by the subcontractor.
3. Repairing the subcontractor's defective or poor workmanship.
4. Completing the subcontractor's work they neglected to finish.
5. Damages for late completion which should be in terms of the agreed Contract Document.

  Sometimes these costs can be significant. Subcontractors should be timeously warned that they will be charged for services and what the costs are. When necessary, they should be given sufficient written notice to rectify problems. Where possible, time sheets for equipment supplied to the subcontractor should be signed daily by the subcontractor, and the back-charges should be agreed and invoiced monthly so that the subcontractor is fully aware of their liability.

  Failure to notify the subcontractor, or get them to sign for the services received, often results in disputes, even ending with the contractor having to withdraw their invoices.

Insurance Claims

  Some events are claimable under either the contractor's or the employer's insurance policies. These include losses due to accidents, theft or from weather damage. Most polices only pay for the cost of removing debris and replacing the damaged items or sections of work. They don't recompense the contractor for lost time, or any penalties or damages that the employer may impose against the contractor for late completion caused by these events. They also won't pay any consequential damages.

Claims for non-payment of invoices

  Late or non-payment by the employer for interim and final valuations and invoices is a grave risk to contractors. It seriously impacts the contractor's cash flow, even sometimes resulting in contractors becoming bankrupt.

It's important that contractors submit their valuation claims in the required format, to the correct person, and to the correct address. Failure to do so could mean the valuation isn't processed on time.

However, some employers make a habit of paying their contractors late. Contractors must monitor payments to check that they are received on time and continually remind their employer when payments are due.

Non-payment of a valuation within the time specified in the Contract Document may give the contractor reason to Terminate the Contract. However, termination can be a lengthy process and seldom ends well for either party. But, having said this if the employer is in financial difficulty termination may be the best outcome to avoid further financial risks to the contractor.

Aside from termination most Contract Documents don't provide the contractor with many other choices when they aren't paid except to claim the interest on the money that's paid late. Some Contracts prescribe the rate of interest that can be charged for late payment. The value of the interest money charged is calculated by multiplying the amount of money due with the number of days it was late multiplied by the prescribed daily interest rate (which is normally the statutory rate in that country plus a percentage point or two).

Most employers don't like to see interest claims for late payments but it may be the only way to correct a serial defaulter. Although it should be noted that some Contract Documents even remove this remedy from the contractor and specifically state that interest can't be charged for late payments.

It's important the contractor understands the country's legislation because in some countries it's possible to lodge a Claim for non-payment of an invoice through the courts which then enforce payment. It should also be noted that many Subcontract Documents contain a clause which says the subcontractor will only be paid when the General Contractor, or Main Contractor, is paid. In many countries this isn't legal and the Subcontractor's invoice must be paid within the Contract's prescribed time period irrespective of whether the General Contractor has been paid or not.

It's important to note that the contractor can't simply stop work because they haven't been paid since they'll then be in Breach of the Contract which would give the employer the right to Terminate the Contract and claim damages from the contractor.

Cardinal Change

A Cardinal Change is when a party makes a change which is so drastic that it requires the contractor to perform duties materially different from those in the

Contract Document. A Cardinal Change is so profound that it's not claimable under the current Contract. The contractor could refuse to do the work and hold the employer in Breach of the Contract. However, care must be taken to get legal advice before proceeding down this route.

Contract Termination

Terminating a contract is fraught with hazards and needs to be done with caution, carefully following the notification procedures as laid out in the Contract Document.

Contract Documents may include a provision for one party – normally the party drawing up the document, which in most cases is the employer – to Terminate for Convenience. This basically gives the party the right to Terminate the Contract without the other party having caused, or provided reason for, the Contract to be terminated. However, many employers erroneously believe that this clause gives them the opportunity to Terminate the Contract for whatever reason they see fit. This is incorrect, and a party cannot Terminate because:

1. They want to appoint another cheaper contractor.
2. They want to remove part of the Scope (even if the contractor isn't performing) and place it with another contractor.

Terminating for Convenience can only be done for rational and honest reasons. So, Terminating for Convenience may be used if for instance the employer is unable to obtain finance, or if they are unable to purchase the land for the project, or cannot obtain the necessary approvals. In some instances Termination for Convenience clauses may be inserted in Contract Documents where the employer has to obtain a prescribed number of sales before construction can begin, such as for an apartment building. Nevertheless, in all cases the other party should be entitled to compensation for proven costs incurred before Termination, unless there is an express exclusion in the Contract Document.

Termination due to Frustration may happen when due to no fault of either party work cannot proceed on the project because it is either impossible or illegal to continue work. It may also arise when the work has changed radically from that envisaged in the Contract Document. Termination due to Frustration must be by mutual agreement and all costs incurred by the contractor to date, as well as those to secure the project and demobilise their resources from the project must be recompensed. Termination due to Frustration may occur when unforeseen ground conditions are encountered which require radically different construction methods, or if the works are flooded for an extended period of time.

Repudiation occurs when one of the parties has breached the contract in a sufficiently serious way that entitles the innocent party to Terminate the Contract and to sue for Breach of Contract. Reasons for Repudiation include:
1. Refusal by the contractor to carry out work.
2. Abandonment of site.
3. The employer employing another contractor to carry out the works.

Wrongful Termination by Repudiation may give the other party the right to Terminate the Contract. It's important to note that the other party has to accept the Termination and that the party Terminating the Contract still has to fulfil their obligations under the Contract. If one of the parties has not accepted the Termination the Contract remains in force. If both parties accept Termination the innocent party is entitled to claim damages from the other party to put them in the position as if the Contract had been completed.

Contractual Termination is for a reason specified in the Contract as cause for Termination. These deal with specific Breaches of one of the parties' obligations.

Some contractors view the Termination clause as an escape clause when they aren't paid, or if they're losing money. Terminating in these circumstances may in fact result in them being in Breach of the Contract, allowing their employer to Terminate and seek damages from the contractor for wrongful Termination.

Before Terminating a contract expert legal opinion should be sought.

Suspension of the Contract

Contracts can only be Suspended if there is a Suspension clause in the Contract Document. Reasons for Suspension are typically the same as for Termination. Suspension allows 'breathing space' for one or both parties to resolve issues between the parties, or to deal with outside constraints on the project. So for instance, if the employer hasn't paid the contractor the contractor may, depending on the Contract conditions, be able to Suspend their work until such time as they are paid. So too, if the employer is waiting for construction permits they could elect to Suspend the works until such time as these permits are received. If the issues cannot be resolved the Contract could eventually be Terminated.

Again the party suspending the works needs to follow all notification procedures as outlined in the Contract Document. The difficulty with Suspension is it cannot be for an indefinite period. Ideally the Contract Document should specify a maximum period, otherwise, the time period will have to be mutually agreed. Failure to have a time limit for the Suspension could lead to problems later.

For example: I was recently told of a project where the client Suspended a portion of the work because they had insufficient funds. There was no time limit set for the Suspension. The rest of the project continued but suffered numerous delays resulting in the contractor still working on the project three years later. The employer then decided they had sufficient funds to continue with the portion of the Contract that had been Suspended so removed the Suspension. The contractor is now placed in the invidious position of having to undertake work at the original Contract rates. In this particular case the contractor may be able to challenge the terms of the Suspension.

It's therefore important that the contractor understands the terms of the Suspension, and both the parties' rights and obligations when the Suspension is lifted, before agreeing to the Suspension of the Contract or a portion of the work.

Wrongful Termination

As mentioned previously when the contract is wrongfully Terminated by one of the parties the other party can sue the party for Breach of Contract and Terminate the Contract. As with any Termination it is important to ensure that it's done within the terms of the Contract and that all notifications are carefully followed.

Use of the Omission clause

In some Contract Documents there is an Omission clause which allows the employer to vary the amount of work in the Contract. Some employers attempt to use this clause to remove work from a contractor for their non-performance, or to give the work to another contractor for reason of convenience, or because the other contractor is cheaper. Removal of work can only be done if the work is no longer required, the employer is removing the work due to budget constraints, or it has so radically changed in design and construction that the contractor will not have the required skills to construct the works.

Time at Large

The term Time at Large is usually used in construction Contracts in the situation where Liquidated Damages could be applied. If Time is at Large then it is argued Liquidated Damages cannot be applied against the contractor because there is no date fixed for when damages will be applicable. However it also means that there is no end date for the Contract. In effect it means the Construction Schedule doesn't have to be adhered to by either party.

Time at Large can occur when there is no agreed Contract Schedule. This may be because work on the project has started but the project is still being developed and the final Scope hasn't been agreed. It could also arise when the contractor and employer can't agree the Contract Schedule.

In some cases contractors apply for an Extension of Time due to events and delays created by the employer. The employer either doesn't reply to the Extension of Time Claim, they refuse to grant the Extension of Time, or there's a delay approving the Extension of Time. With no Extension of Time granted the contractor could face the risk of Liquidated Damages being applied for late completion. In this case the contractor may say Time is at Large.

Even if Time is at Large the contractor still has the duty to complete the works as quickly as possible. Declaring Time at Large is not an excuse to work more slowly.

It should be noted that Time at Large is not generally considered a legal term and it could be tested in court. Contractors sometimes send a letter to the employer declaring Time at Large and think their problems will be solved. This might not be the case and legal advice should be sought if the contractor is hoping to use this as a reason to avoid Liquidated Damages.

Liquidated Damages

Liquidated Damages are usually a designated amount predetermined in the Contract Document to compensate for a specified Breach to the Contract – usually late completion. This amount is to compensate the injured party and not to punish the offending party. The amount of damages must be reasonable and should be provable to some extent by calculations. They should represent the best estimate at the time of drawing up the Contract. So for instance if the Liquidated Damages are $1,000 per day the employer should be able to substantiate the amount by saying that was the amount of rent they had to pay to remain longer in their rental premises.

A Penalty is a sum which is disproportionate to the value of the damages suffered. A Penalty is unenforceable in many countries.

Thus, if the project is completed late, but the contractor can prove that the employer didn't suffer a loss, or that the amount of Liquidated Damages was disproportionate to the actual costs the employer incurred because of the late completion, the contractor would be able to dispute the imposition of the Liquidated Damages. So for example; General Contractors who try to impose

Liquidated Damages on their subcontractors may face difficulties if their employer didn't enforce Liquidated Damages on them.

For Liquidated Damages to be enforceable they should be the exclusive remedy available to claim for the Breach. In other words the employer cannot claim Liquidated Damages as well as other damages for the same Breach.

It should be noted that Liquidated Damages in the Contract Document serves to protect both the employer and the contractor. If there is no Liquidated Damages clause, or if it is deleted, marked 'Nil' or left blank then the offending party may be liable for the full cost of damages suffered by the other party. These losses could even include consequential losses such as loss of business.

Claims by subcontractors against contractors

It's important for contractors to note that subcontractors can lodge Claims against them for the delays and Variations outlined in this chapter. This includes circumstances where the contractor has delayed the subcontractor, when the contractor hasn't fulfilled their contractual obligations as outlined in the subcontract agreement, or where the subcontractor has incurred additional costs due to changes or additions caused by the contractor. Contractors should take mitigating steps to ensure their subcontractors have no reason to lodge Variation Claims against them.

When the Claim is as a result of the employer's actions then the Claim needs to be passed onto the employer within the specified time for the submission of Claims. Contractors need to deal with the subcontractors' Claims swiftly and in accordance with the Contract with the subcontractor to avoid disputes.

Claims by the employer against the contractor

As discussed previously the employer can apply Liquidated Damages against the contractor for not completing the project in accordance with the Construction Schedule. They can also Terminate the Contract for Breach of Contract by the contractor.

There may be other Claims and remedies that they could lodge against the contractor depending on the Contract Document. These could include:
1. Failure of the contractor to deliver a product that meets the required quality standards or satisfies the specifications. This could be resolved by:
    a. The contractor repairing or replacing the structure so it does comply.

b. If the contractor fails to rectify the problem, and after due notification, the employer could engage another contractor to carry out the remedial work and deduct the actual cost of the repairs from the contractor's account.
c. In some case the remedial work could be impractical, or cause the employer more harm, even delaying the project, especially if the offending structure has to be demolished. The employer and contractor could come to a mutually acceptable agreement where the employer accepts the non-complying item on condition the contractor agrees to be reimbursed a lesser amount than if the item had conformed.

It should be noted that the employer cannot unilaterally deduct money from the contractor for non-conforming work. The contractor needs to be given written notification that the work doesn't conform and the reasons it doesn't conform, and then given a time by when it must be rectified. Only if the repairs aren't completed in the specified time can the employer engage another contractor to repair the defective work and pass these costs onto the contractor. The exceptions to this may be when the failure of the structure poses a direct threat to peoples' safety or when the damage stops a critical operation of the employer, in which case the employer is entitled to carry out immediate remedial work while at the same time notifying the contractor of the problem and remedies taken.

2. Back-charging the contractor for services, material, equipment or people which the contractor was obligated to supply but which the employer or the employer's contractors supplied. However, employers need to ensure that these are documented and agreed so there aren't disputes over the quantities, hours and rates later. In some circumstances employers have been known to 'force' additional people or equipment onto the contractor when they decided (unilaterally) that the contractor needed additional help because they were falling behind the Schedule. In this case the contractor could dispute the charges for these resources claiming that they weren't required nor requested.

3. Damages for various Breaches outlined in the Contract
    a. Non-compliance with confidentiality clauses.
    b. Failure to comply with procurement requirements.
        For example: the employer may have specified in the Contract Document a minimum value (or percentage) of items that

should be procured from local providers or prescribed disadvantaged groups.
   c. Not complying with the employer's security provisions.

The amount of Damages should be included in the Contract Document otherwise the employer needs to prove that they incurred the costs they are claiming as Damages against the contractor.

4. The actual costs the employer incurred because of the actions of the contractor. These costs could be for:
   a. The employer's personnel having to work longer hours to accommodate the contractor's work hours and methods where these are unreasonable or could not have been foreseen by the employer.
   b. The costs of witnessing the contractor's tests that failed and had to be redone.
   c. Costs for the design team to redesign structures to comply with the contractor's chosen methods of construction which weren't included in the contractor's tender submission. Also costs for the design team to confirm whether construction errors could be accepted without compromising the design and integrity of the facility.

Summary

1. Contractors can have reason to lodge a Claim if they incur a delay or increased costs due to reasons beyond their control providing these reasons haven't been excluded in the Contract.
2. Contractors could incur additional costs because of a delay, if their work methods are disrupted, if the Scope of Work is increased, specifications are changed, the Contract is altered, the employer hasn't fulfilled their obligations in terms of the Contract or if they have to Accelerate their work due to reasons which aren't caused by them.
3. It is important contractors understand the Contract so they know if they can submit a Claim for an event. Not all events are claimable.
4. Just as the contractor can lodge a Claim against the employer or their subcontractors so can subcontractors and the employer lodge a Claim against the contractor for delays and additional costs they incurred as a result of changes and delays caused by the contractor, or because the contractor didn't fulfil their obligations under the Contract

5. In some cases Claims could lead to Suspension of the Contract or even Termination. However, contractors need to understand the Termination clauses because wrongful Termination of a Contract could give the employer (or subcontractor) cause to lodge a Claim against the contractor.

Chapter 2 – Supporting Documentation

Contractors think that putting a Variation Claim together starts with preparing the Variation Claim letter. Unfortunately by this stage if the contractor hasn't put some basics in place their Claim may be doomed to failure.

The Contract Document

The Contract Document defines the relationship between the parties on a construction project and forms the basis of all Claims. The documents include the Contract, General and Supplementary Conditions, the Construction Schedule, drawings and specifications. Misunderstanding or misinterpreting these documents often leads to disputes.

Regrettably many Construction Documents are one-sided and drawn up to favour the employer. The Contract may place undue risks on the contractor, risks which might in the normal course of projects be risks that the employer should accept – risks that the contractor can't control. So for instance, if the Contract Document doesn't allow the contractor to claim for certain changes, or lodge Extension of Time Claims for some events, the contractor's Claim for these changes or delay events won't be accepted.

In addition the Contract normally specifies a maximum time by when the contractor must notify the employer of their intention to submit a Claim. Sometimes these time limits can be onerous. Failure to submit a Claim within the time may result in it being Time-Barred, in which case the Claim can be rejected.

Often employers don't issue the Contract Document until well into the project. This leaves the contractor exposed should things go wrong. The Contract Document is there to protect both the contractor and the employer.

Case study: One of my projects was the civil works for a substation where we were a subcontractor to a main contractor. From the start the main contractor delayed us by issuing construction drawings and information late. In addition, we encountered extensive rock on the site which was excluded in our price. Furthermore, the construction drawings differed from the tender drawings, and they were changed as work progressed resulting in us having to redo completed work. We documented all the delays and changes, and notified the main contractor via appropriate correspondence as they occurred.

As happens with many projects, we were issued a Letter of Intent to start the project, but were only issued the Contract Document two months after we had begun work on the project. Without making any changes we signed and returned the Contract Document two weeks after receiving it. The Contract Document included the Construction Schedule which was submitted at the start of the project before the delays were incurred.

Throughout the project the main contractor didn't respond to our Variations or Claims for Extension of Time. When we finally got the contractor to sit around the table to resolve all the Variations and Extension of Time Claims we had lodged they refuted the Extension of Time Claims on the basis that we had signed the Contract Document, which included the original approved Construction Schedule, after most of the delay events had occurred. By doing so, we had inadvertently accepted the original Schedule as though it included the delays which occurred between the project start date and the date the Contract was signed.

The lesson from the above is contractors shouldn't start a project before the Contract Document has been agreed and signed. If, for some unavoidable reason, this doesn't happen, then the contractor must ensure the Contract eventually signed takes into account all changes, Variations, delays and new information which occurred up to the date of signing. Alternatively the contractor could include a letter stating that the Contract Document signed only takes into account information that was known at the start of the project.

Before signing the Contract Document the contractor should ensure that the document is the same as the one they priced, and that any addendums, changes and conditions included in their price, or negotiated afterwards with the employer, have been incorporated into the Contract Document. It's important to check particular clauses since even minor changes in wording or punctuation can significantly change the conditions of the Contract. Also, the contractor should check that the drawings included in the Contract Document are the ones they priced. Some employers include additional drawings in the Contract Document, or drawings that have been revised after the contractor priced the project. The Contract Document is the document against which all Variations and Extension of Time Claims will be measured and should be checked carefully by the contractor before signing and agreeing to its terms.

Conflicts between clauses, or clauses that are ambiguous should be resolved and corrected before the Contract is signed.

But more important is for the contractor's Project Manager and the contract management team to read and understand the Contract Document. They must understand their obligations as well as those of the employer. They need to know who correspondence must be addressed to and by when Variations Claims must be lodged.

I've found, particularly when using a new form of Contract, it can be useful to arrange for the contractor's Project Manager and their contract team to attend a workshop explaining how the Contract should be executed. This is a small cost which has more than paid for itself on our projects. I've also arranged a workshop for both our team and the employer's team which proved useful and aligned both parties, thus avoiding potential misinterpretation of the Contract Document later.

Tender drawings (Drawings issued with the request for price)

The drawings issued by the employer with their request to price are valuable as they normally form the basis of the contractor's price. Any deviations from these drawings usually provide reasons for the contractor to lodge a Variation Claim.

I normally make at least two copies of these drawings and ensure one copy remains with the original tender documentation and another copy is clearly marked 'Tender Drawings' and is close at hand on the project site. As new Construction Drawings are issued these are compared to the original tender drawings to check how the drawings differ.

These drawings should have been included or referred to in the Contract Document.

The Construction Schedule (Programme)

The Construction Schedule must be approved in writing by the employer. The Construction Schedule can also be approved in the minutes of a project meeting. Unfortunately many employers are reluctant to approve the Construction Schedule in writing or continually refuse to accept the Schedule. As long as the Schedule isn't approved the contractor has difficulty proving the employer delayed the project – how can the delay be quantified without an approved Schedule?

The Construction Schedule should include all the tasks, their durations and the links between the tasks. It should show when access is required to the work areas.

It's important that tasks on the Schedule are correctly linked. Sometimes contractors don't link adjacent structures even though the construction of one may impact the construction of the adjacent structure. The adjacent structures may impact access, or the one may for instance be founded deeper, necessitating it to be done first. It's particularly important when we have utility services (such as sewers, water pipes, stormwater drains and gas mains) crossing the construction site that these are linked to the structures they impact, or which impact their installation – sometimes these utilities pass under structures so should be installed before the structures are constructed. Incorrect links may mean that when one task is delayed its impact on the adjacent structures isn't properly recorded.

Having a fully resourced Schedule which is provided to the employer can be valuable when negotiating Acceleration Claims.

It should be noted that even when a Construction Schedule has been approved by the employer it could be rejected when determining an Extension of Time Claim if:
1. There's a technical or logic error which makes the Schedule unworkable.
2. The Schedule hasn't been followed in the construction process or if different construction methods have been utilised from those shown in the Schedule.
3. The Schedule didn't take into account known restraints.
4. No allowance has been made in the Schedule for normal adverse weather which would impact progress.
5. There are unrealistic durations.
6. The Schedule is incomplete.
7. The Schedule isn't in sufficient detail to allow the calculation of the delay impact.
8. The Schedule is impractical.

It's useful to issue the employer with an Information Required Schedule (log or register) which indicates when the employer must issue drawings and information to the contractor. This schedule is a product of the Construction Schedule and takes into account the procurement Lead Times.

If the Construction Schedule is revised during the course of the project because of an approved Variation, or for some other reason, the new Construction Schedule must again be approved in writing and a revised Information Required Schedule should be issued to the employer.

The Construction Schedule is the basis of all Extension of Time Claims so it's important it is correctly prepared and that it's regularly and accurately updated. It should be modified in the course of construction to take into account Variations, delays and changes in sequencing which have been agreed between the contractor and the employer.

Schedule Float

Float is the amount of time that an item can be delayed before it impacts on the Completion Date of that section of works, or the project as a whole. Float can be an emotive subject as the employer normally believes the Schedule Float belongs to them and any delays they caused that reduce this Float (but which don't impact the end date of the project) aren't grounds for an Extension of Time Variation. The contractor, on the other hand, believes the Float belongs to them, and any delay by the employer which impacts and reduces it is reason for a Variation, while the delays caused by the contractor should first reduce the Float before impacting the project duration.

Often the employer makes it clear in the Contract Document that they consider all Float to belong to them. In these cases as a contractor I would always try and ensure that where possible the Float in the Schedule is hidden and any visible Float is kept to the minimum.

At tender stage the contractor could also try to ensure that it is clearly stated that the Float belongs to them.

Where there is no clear statement as to who owns the Float, the norm is that it is used by whichever party requires it first. In other words, whoever causes the delay first will use the Float to cover that delay. Only once the entire available Float has been used will a claimable delay become an Extension of Time Claim.

For example: if there is 10 days Float on the Critical Path of the Schedule and the project is delayed by 2 days because of adverse weather there would be no Extension of Time Claim and the Float would be reduced to 8 days and the Completion Date would remain the same. If the contractor then caused a delay of 5 days because their materials arrived late there would again be no change to the Completion Date but the Float would be reduced by 5 days to become 3 days. If after all these events had occurred the contractor unexpectedly encountered a buried gas pipe which stopped further work on the project for 5 days while it was relocated, then the contractor would be able to claim an Extension of Time of 2 days. The first 3 days of this 5 day delay is taken up with the remaining Float.

However, if in this example the Float was all claimed by the employer then the delay of 5 days caused by the contractor's material arriving late would not have reduced the Float. Consequently the delays caused by inclement weather and the gas pipe would have been a total of 7 days which would have been less than the 10 days of Float and no Extension of Time would be granted.

If the Float belonged to the contractor they would be able to claim for both the weather delay as well as the gas pipe delay and could claim a 7 day Extension of Time.

Resource driven Schedules

Most Construction Schedules are logic driven with links between activities dictated by what has to be completed first before the next activity or task can follow. Logic driven Schedules may result in excessive peaks and troughs in the contractor's resources, with sometimes resources not being required and then the next day many are required. To overcome this problem contractors try and schedule tasks to ensure continuity of their resources and in other cases extend the duration of some activities so that fewer resources are needed to complete the task. In most cases employers are only interested in logic and want their project completed as quickly as possible.

When contractors prepare Schedules around their available resources, making the best use of their resources and smoothing out peaks and troughs the Schedule becomes resource driven. In some cases the contractor plans not to start a task until the required resources become available from another task. Logic may dictate the task can begin earlier because the preceding task it's linked to is completed, but the contractor requires resources to become available from another task which isn't necessarily the preceding task dictated by logic. It is essential contractors include these resource driven links in their Construction Schedule and that the employer is made aware that the Construction Schedule is not only based on logic but also on the efficient use of the contractor's resources.

All too often I've experienced employer caused delays which have negatively impacted my carefully thought through Construction Schedule which was based on both logic as well as the best and most efficient use of our resources, thus disrupting my resource utilisation.

Updating the Construction Schedule

Employers normally require their contractors to regularly update the Construction Schedule. These updates are often included in the project meetings and in project progress reports. Unfortunately some contractors don't update the Construction Schedule correctly and show their actual progress incorrectly. The updated Schedule may be used when assessing the contractor's Claim for

Extension of Time. If the employer issued information late, but the updated Schedule showed the contractor behind, then it could be judged that there was a Concurrent Delay with one delay caused by the employer and the other by the contractor.

Just as easily the updated Schedule may incorrectly show progress on an item when in fact there wasn't progress because the contractor was waiting for information to start the task. The updated Construction Schedule would therefore be in conflict with the Extension of Time Claim the contractor submits.

The updated Construction Schedule must be completed correctly, physically checking all tasks.

Documentation

A Variation Claim can be more easily supported if the contractor has maintained accurate documentation and records during the course of the construction project, communicating regularly with the employer, and has taken all reasonable steps to prevent the Claim from arising.

It is important that documents are distributed in a timely fashion in accordance with the Contract. A delay in distributing documents could result in a delay to the project, or it could prevent one of the parties from taking timely action to prevent a Variation or Extension of Time Claim from arising.

Documents must be distributed to the appropriate person. Contract Documents usually specify who that person is. If in doubt it's pertinent to check who should receive the different types of documents. If the appropriate person doesn't receive the document then it is the same as if the document wasn't issued. In particular Contract Notices have to be delivered to a particular person at a specified address and failure to adhere to these instructions may make them null and void.

Even within the contractor's project team documents need to be distributed in a timely fashion to the correct people so that delays do not occur.

> *For example: the contractor could receive construction drawings from the employer on time, but work could be delayed if the drawings aren't immediately distributed to the person who has to procure the materials required for the work indicated on the drawing, or if the Supervisor who has to do the work doesn't receive the drawings on time. Delays caused by the contractor's own internal system are obviously not claimable.*
>
> *Equally important is to ensure that designated representatives of the subcontractors also receive information immediately after the contractor receives it.*

It is imperative that the employer and subcontractors are immediately advised of organisational changes in the contractor's project team. These changes could change who the responsible person is on the project, or change the person responsible for a particular section of works. Failure to provide the correct notifications could result in document transmissions going astray or being delayed.

Documents should be filed where they are easily retrievable so that they can be readily referred to. Electronic documentation should be stored on a central storage system. More than one Claim has been jeopardised by the theft of a computer which has resulted in the loss of valuable supporting documentation.

Even draft Claims should be stored where they are secure. Preparing a Claim often requires days and even months of work, which could all be for naught if the work is lost.

Documentation should be securely filed. Many contractors are negligent and leave confidential paper documents lying on tables where they can be viewed, or even stolen. Electronic documents should be password protected so unauthorised access isn't possible. Systems should be secure so they can't easily be hacked. The loss of sensitive or critical documents could jeopardise the success of a Claim. Should the employer, or a subcontractor, involved in a Claim with the contractor come into possession of confidential documentation relevant to the Claim it may assist them to better defend the Claim. Also, should the employer see the contractor's cost reports or budgets which indicate that the contractor will be profitable on the project they may be less sympathetic to a Claim from the contractor.

Instructions

It's essential that the employer issues written Instructions when these instructions could result in a Variation. All too often employers issue verbal Instructions which may be misinterpreted or later denied by the employer.

Instructions must be signed by the employer's designated representative who is authorised to issue Instructions. They must also be authorised to issue Instructions of that value, since sometimes some representatives have a limit on their authority, and exceeding that limit may mean the Instruction isn't valid.

Only the contractor's authorised representatives should sign receipt of the Instruction to ensure that it's checked properly. The contractor must understand the intent of the Instruction since it may only be intended for the contractor to price the additional work, or changed scope, and work shouldn't proceed until the employer approves the contractor's price and issues a formal Variation Order.

Anyway, it's normally best not to proceed with the Variation work until the contractor has priced the Variation and the price has been approved. Proceeding with the work could leave the contractor exposed to the risk of the employer not agreeing to the contractor's price. Unfortunately most construction projects are fast-track and contractors seldom have the luxury of following this process.

It also pays to check the wording of Instructions as I've found that some wording may imply the Instruction is a result of the contractor having to redo work or not complying in some way with the drawings or specifications, which could mean that the employer doesn't have to pay for the work included on the Instruction. In addition I've had Instructions state in the small print at the bottom that there is no cost to the Instruction. It should be noted that often agreeing the cost of these Instructions is in the hands of Contract Administrators and Quantity Surveyors who might not be familiar with the operations on the project and can only interpret the Instruction as they read it. If the Instruction was ever disputed the third party dealing with the dispute would interpret the Instruction as it reads. Therefore, the wording on instructions must be clear and unambiguous.

Contractors need to check the wording on Instructions to ensure that they can comply with the Instruction. Employers sometimes insert dates by when the work must be completed which might be impossible to meet, or carrying out the work on the Instruction could disrupt the existing Construction Schedule.

Weather records

To be able to claim for delays caused by inclement weather the contractor needs to be able to record the actual weather from the start of their work to prove it's in excess of what could normally be expected. It's good practice to install a weather station on the project to record rain, maximum and minimum temperatures and wind speeds. At the very least install a rain gauge. The measurements should be included in the Daily Log or Diary and the employer should have the opportunity to verify the accuracy of the recordings. On project sites that cover a large area it may even be necessary to install more than one weather station to take account of weather variations that can occur across the project site.

Typically the contractor can only claim for weather events in excess of the average conditions for that area in that month. Average weather records are usually available from the weather station closest to the project. However, care should be taken that these records are applicable to the project site as weather conditions can vary dramatically over even small distances.

Project meeting minutes

All projects should have regular project meetings to discuss progress and problems on the project. It's essential that accurate minutes are kept of these meetings as they can form an important document should a Variation Claim arise. Often employers write up these minutes and they can ignore relevant facts, or misrepresent what was discussed at the meeting. Sometimes these omissions are an oversight but often it's so they aren't put in a poor light. The onus is on the contractor to check that the minutes are an accurate reflection of what was discussed at the meeting, and if they aren't that they ensure the minutes are corrected or that they record in writing, to the employer, the necessary corrections.

Items in the minutes of these meetings that require the contractor's attention or response should be dealt with immediately. Items that aren't closed-out and which remain on meeting minutes for several meetings may give the impression that the contractor isn't responsive to the employer's requests. This may be perceived negatively by outsiders adjudicating Claims.

Daily Logs or Diaries

Daily Logs or Diaries should be prepared by the contractor as the project progresses. The Daily Log should record:
1. The date.
2. The weather conditions.
3. Manning, including subcontractors and day visitors.
4. Major items of equipment.
5. Delays.
6. Significant tasks worked on or completed.
7. Major items of material received.
8. Safety, environmental or industrial relations incidents.
9. Potential future delays.
10. Any other relevant information.

Preferably they should be signed by the employer or there representative and a copy submitted to the employer. Daily Logs cannot be prepared after the fact. The Daily Logs should be consistent, since inconsistencies and discrepancies cast doubts over the reliability of the information recorded and used to support a Claim.

Frequently the task of these reports is left to a junior member of the construction team who might not have all the facts or may not realise how important it is to ensure that these reports are accurate. Unfortunately resources aren't always correctly recorded and the subcontractors' resources, or the construction team in the office, are overlooked. Manning and equipment numbers can become significant when delays occur, when Claims for loss of productivity arise, to formulate Variation costs and for Acceleration Claims. Manning and other information recorded in the Daily Logs should be the same as that recorded in the meeting minutes and project reports.

Where possible equipment breakdowns should be recorded as well as what tasks the equipment was working on.

In addition the Daily Logs or Diaries should record all delays. Even when the delay was recorded the day before, if it's ongoing, it should be recorded again as a delay. I have had contractors record the delay when it first happened and then not record it again in the Daily Log even though it was only resolved several days later. Unfortunately employers reading the Daily Log will only see the delay as being recorded on one day and it will be difficult to prove it lasted longer.

Contractors should insist that their subcontractors submit a Daily Diary which the contractor should check before signing. Since this is an important task, and the document could form supporting evidence to the subcontractor's Claim later, the accuracy of the Diary must be verified by a competent person who understands the importance of this task and has knowledge of the work the subcontractor is performing.

Personal Diaries

It's good practice for Project Managers and Supervisors to keep their own diaries of events that occurred on the project. These could include discussions they had with subcontractors or the employer's team, concerns, progress, and equipment breakdown. These diaries could be invaluable references to refer to when formulating and defending Claims, providing accurate date records of events.

Although these may appear to be personal diaries it should be noted that in the event of a dispute the other party may be entitled to subpoena individuals' diaries so it is important that these documents are accurate and remain factual.

The diary entries should be dated on consecutive pages in a bound document. Loose leaf pages could mean that the diary has been written after the event.

Request for Information (RFI) or Engineering Queries.

From time to time the employer's construction drawings aren't clear, have discrepancies, contain conflicting information, or the information is insufficient for

construction to proceed. It's then necessary to ask the employer for clarity, or for the missing information. Often contractors do this verbally and there is no record that the question needed to be raised, when the question was asked, and often no record of what the answer is. It's therefore important to ensure the question and the answers are recorded in writing. On our projects we always had 'Question and Answer' books. Elsewhere these may be known as RFI's (Request for Information) or Engineering Queries. Each question should have:
1. A specific sequential number so it can be referenced.
2. The date the question was asked.
3. The name of the person asking the question.
4. Detail of the question including referencing drawing numbers where necessary.

Preferably the question should have a space for the answer to be added as well as the date when the question was answered and the name of the person who provided the answer.

A register should be kept of the RFI's to track how many have not been resolved. This register is also useful for formulating Claims because:
1. The time the employer takes to answer the question might give cause for an Extension of Time Claim.
2. The actual answer may require additional work or entail extra costs which could be reason for a Variation Claim.

Care must be taken to check that the employer has actually answered the question asked, and that the answer is one that can be used. Often employers may misunderstand the question so the answer provided is incorrect in which case the question has to be rephrased and resubmitted. Frequently employers answer with the phrase; "drawings to follow", which obviously isn't a useful answer and the question on the register can only be closed out when the relevant drawing which satisfactorily answers the question is received.

Each RFI should include only one question. Often questions relate to different disciplines such as; electrical, architectural, or mechanical, so different people in the design team have to answer the questions. Having more than one question on the RFI could delay answers to questions, or result in some questions being missed, resulting in possible delays to the project. RFI's must be directed through the appropriate channels set up by the employer to ensure they aren't misplaced and that they reach the correct person. Questions from subcontractors, as well as the answers, must be directed through the contractor so they can be tracked.

Numerous RFI's are often an indication that there are problems with the construction drawings and information issued, and the contractor may need to discuss the quality of the drawings with the design team.

Emails

Emails normally form part of the legal correspondence on projects. Unfortunately many emails are treated as casual correspondence with some contractors not paying enough attention to the content of their emails or their replies to the employer's emails.

For example: In response to a delay or extra work the contractor's Project Manager may be quick to send the employer an email reply with a simple statement like; "no problem", or 'that shouldn't be a problem", or "there shouldn't be a cost", or something similar. Employers may then be surprised when they later receive a formal Claim from the contractor. Employers could claim that they were misled and if the contractor's email had mentioned there would be costs involved they would have considered alternative solutions.

Another possible problem with emails is they are often copied to multiple people. It's easy for an email, with sensitive information, to be forwarded to the employer by mistake. Contractors wouldn't want a Claim to be sunk before it was even submitted because the employer saw an internal email regarding the Claim with differing information to the final Claim they received.

It's easy to quickly write an email that's emotional and even possibly slanderous. Keep all communications civil. If the contractor wrote an email which called the employer or one of their team various names it could be used against the contractor if a Claim went to arbitration or litigation.

Letters

Letters could form important supporting documentation for the contractor's Variation Claims. Letters should:
1. Be addressed to the correct representative as named in the Contract Document and copied to other necessary parties.
2. Have a date.
3. Include a reference number.
4. Have a heading, including the project's name and reference number, and then the subject matter.
5. Be concise in simple language.
6. Not use emotive language.
7. Quote the correct Contract clauses and reference drawing numbers where applicable.
8. If they are a reply to another letter, reference that letter.

9. Clearly summarise the facts, and where necessary the action or response required

It's often good practice to maintain a log or schedule of letters received and sent. Also, ensure where necessary that employer's letters which require a response are responded to. However, caution should be taken not to enter a petty letter writing war over trivial matters.

Photographs

Photographs with a date are essential on all construction projects. They can play a valuable part in providing supporting evidence for Claims submissions as well as for refuting Claims lodged by the employer or subcontractors. They can be used to record:
1. The progress of the work – it's useful to take photographs from the same location each week.
2. Access problems caused by the employer or other delay events resulting from their actions.
3. Completed work that has to be redone because of revised drawings or new Instructions.
4. The number and type of equipment working on a Variation.
5. Claimable events such as excavating in rock where none was allowed, or working around previously unknown existing services or other obstructions. (Photographs showing depths, levels or size may require a tape measure or another object to gauge the size and dimensions.)
6. Damaged caused to the works by a subcontractor, the employer or another contractor.
7. Damage caused by an insurable event, such as; flood or wind damage.
8. If the employer isn't available to inspect the works before it is covered over, to verify that a particular task or item has been attended to, or to record the quality of an item before it is covered up.
9. Poor quality work executed by subcontractors which requires rectification.
10. How a particular task was executed so those pricing the Variation can assess how this differed from what was anticipated in the Contract Document.
11. Damaged equipment or materials delivered to the project.
12. Subcontractors' progress.
13. In the event that the contractor completes a section of work late for which they may face Liquidated Damages from the employer, to record

the employer's actions following when they were given access. Often employers aren't ready themselves to take access, yet, they are quick to apply Liquidated Damages for the contractor's late completion. Employers can only apply Liquidated Damages if they have suffered financial loss. Recording the employer's lack of action and urgency after handover may help refute their Claim against the contractor.

Photographs must be given a title which includes the project name (or number), the location of the photograph and the reason for taking the photograph. This should be done immediately so there is no confusion. Often when formulating Claims contractors are unable to find all the photographic records of the event because the photographs haven't been given the correct title or aren't filed correctly.

All photographs, even those of poor quality, should be kept, since they may prove useful in the event a Claim arises. Frequently when we had a Claim we found supporting evidence in old photographs that were taken for other reasons, yet they just happened to show progress or some other item relevant to support our Claim, or to refute a Claim made against us. Even a grainy poor quality photograph will be better than no photograph at all – after all this isn't a photographic competition!

Video recordings can include even more information than a photograph, particularly when showing a process that has been negatively impacted due to circumstances which give rise to a Claim.

Drawing Register

A Drawing Register is essential on every construction project and should record when drawings were issue to the contractor. Even when the employer updates and manages the Drawing Register the contractor should check it to ensure that they actually received all the drawings recorded as being issued to them, and that they were received on the day recorded. Drawing Registers are an important document to record when information was received. Even a day's discrepancy can impact an Extension of Time Claim, ultimately perhaps costing the contractor thousands of dollars.

It's also good practice to stamp all construction drawings with the date they were received.

Distribution of drawings

Contractors receive drawings from their employer and then issue them to their subcontractors. Delays in distributing these drawings could delay the project.

It's important that there's a record (agreed by the subcontractors) of what drawings were issued to the subcontractors and when. On many occasions I've had subcontractors claim that they hadn't received a particular drawing or revision, which we could prove they had received, thus refuting their Claim against us. Drawings should be issued with a drawing receipt which records; the drawing number and revision number, drawing title, the date of issue, who issued the drawing and who received the drawing. These receipts should have proof that the drawing was received which could either be an ink signature or electronic verification.

It's important that drawings issued to subcontractors are done through the channels agreed with the subcontractor. Sometimes drawings are issued directly to subcontractor's site personnel and there's no record of it being issued, or the subcontractor's Project Manager is unaware of the drawing being received.

Project Reports

Some employers insist on Project Reports on a weekly or monthly basis. These reports are an opportunity to record progress and report on delays that occurred. It's important that these reports are correct and the information contained in them agrees with the information recorded in the project meeting minutes and in the Daily Logs. Discrepancies can cast doubt on the facts reported and may be detrimental to future Variation Claims. Forgetting to record a delay in the report could also be used against the contractor should a Claim for the delay be disputed.

Searching for evidence of claimable events

As mentioned in Chapter 1 there are many reasons for Variation Claims. However, these are not always obvious to the project team and the contractor needs to regular review a number of documents to check that there isn't a Variation Claim that they've missed. These documents would include:
1. New construction drawings which should be compared with the drawings included in the Contract Document.
2. New revisions of construction drawings to see what has been changed, what's extra, is there work which is completed that has to be redone, or are there items on previous revisions that have already been procured which are no longer required.
3. Site Instructions and Variation Orders issued by the employer.
4. Project Meeting minutes which could contain instructions or reasons for delays.

5. Drawing Registers to confirm information has been issued in accordance with the Construction Schedule.
6. Daily logs or Diaries which may record delays or weather events.
7. The RFI register to confirm that requests for information have been replied to timeously and haven't delayed the project. Numerous RFI's may be cause for a Disruption Claim since some RFI's are generated because of problems on the construction drawings – problems which may cause work to stop until they are resolved.
8. Answers to RFI's which sometimes require the contractor to change a detail on the construction drawing, or may clarify a detail that's not clear on the drawing.
9. Letters and emails which could record delays, or include information and instructions that are Variations.
10. New project specifications or Contract Document amendments which could contain changes which impact the contractor's costs.
11. Quality reports which might contain additional tests or inspections that weren't allowed for in the contractor's pricing. The employer could also be demanding a higher quality or specification than the Contract Document calls for.
12. Shop Drawing approvals to confirm that the employer has returned the drawings in the time frames specified in the Contract Document, and that they haven't requested additions and changes on the drawings which could have a time or cost impact.
13. Punch Lists (Snag Lists) to confirm that all items recorded are attributable to the contractor and that there aren't additional items which weren't required in the Contract Document or items that were damaged by the employer or their other contractors.
14. The updated Construction Schedule including questioning why there's slippage. This could uncover work that is being delayed by the employer, or even an increase in Scope.
15. Minutes of subcontractor's meetings as well as subcontractor's correspondence since there may be evidence of delays, variations and frustrations experienced by the subcontractor, caused by the employer, which could give rise to a Claim.
16. The contractor's project cost reports. If there is a loss find the reason. Sometimes losses could indicate that a change or Variation has occurred on the project which hasn't been claimed. Lower productivities may be a sign of disruption, delays or changes in the construction sequence caused by the employer or other events which aren't a result of the contractor's

actions. It may also indicate that the contractor has supplied services to their subcontractors which should have been back charged to the subcontractor and haven't been.
17. Project permits. When the employer is responsible for obtaining environmental, access or occupation permits it is wise for the contractor to check the conditions on these permits since the contractor may have to comply with conditions which weren't included in the bid documentation and which entailed additional costs for the contractor.

In addition talk to the crews in the field to understand their frustrations and the reasons progress is being held up. Sometimes management doesn't always understand what's happening in the field and how the employer's representatives are interfering or disrupting progress. Talking to the field employees may also highlight other additional problems and obstructions encountered on the project.

Is the contractor's team really on their side?

Contractors may say one thing in meetings with their employer, in correspondence and in project reports, but, is this the same message the employer is hearing from the contractor's personnel in the field? Employers and their representatives frequently interact with the contractor's team working in the field. In fact, communication between employer and contractor is to be encouraged, but when the contractor's employees say the wrong thing, or are critical of their management or company, in conversations with the employer it could wreck the contractor's Variation Claim.

It's important everyone in the contractor's team understands the reason for delays and changes. These delays can cause frustration and often those in the field may be quick to blame their own management for a problem which, unbeknown to them, is due to the employer. It's also important that the employer hears a consistent message about what's causing the delay from the contractor's team. Inconsistencies can cast doubt in the employer's mind as to the validity of the Claim, or even give reason for them to refute the Claim.

The contractor's employees shouldn't be telling the employer about mistakes their management has made or how the contractor is delaying the project, nor be critical of the rest of their team. This will ultimately give the employer ammunition to refute the contractor's Variation Claim.

For example: I submitted a Claim for an Extension of Time due to the employer providing access late. The employer tried to refute this Claim because they heard from one of our Supervisors that we weren't ready to start on that section of works because our resources weren't available.

Records of Variation work

It's important that accurate records of Variation work are maintained. Where possible these records should be agreed with the employer and signed by their representative on a daily basis as work proceeds. Photographs with the date are also valuable evidence.

These records could include:
1. The elevation and quantities of rock or other obstructions, where relevant.
2. Equipment and man hours worked on the Variation. This should also include unproductive times where necessary.
    > *For example: often workers work directly on a task for say six hours. But they were actually paid eight hours for the day. The remaining time was taken up attending the morning safety meeting, preparing hazard assessments for the task, organising equipment, collecting the required materials, accessing the work area and their paid rest breaks. Clearly the contractor who only invoices for six hours is going to be two hours out of pocket.*
3. The exact times of delays – when worked stopped and when it could be restarted.
4. Evidence of work already completed which has to be redone – get the employer to witness and agree the work and take photographs.
5. The materials received which can no longer be used – this could include invoices and delivery receipts.
6. Receipts for materials received where the cost of these materials will form part of the claim.
7. Time and equipment used to offload materials for the task.
8. Tally count of truckloads of material where this is the agreed payment method or when it will be relevant to formulating the Variation costs.
9. Tools and material taken from the contractor's store.

It's also pertinent to document any further delays that impact the Variation work.

Hand-over documentation

Before the employer or their other contractors take access from the contractor of a section of completed work a detailed Snag List or Punch List should be prepared detailing any incomplete work and defects. Without this defects list the contractor isn't able to later prove that other damages were caused by the employer or their contractors, which could result in the contractor fixing these damages at their cost. With an agreed Snag List the contractor will be able to submit a Variation Claim to rectify damages caused by the employer or the employer's other contractors after they occupied the section.

It should be noted that before repairing these damages the contractor should notify the employer that the damages were caused by the employer or the employer's contractors. The employer then has an opportunity to get others to repair the damages, or to issue a Variation Order to the contractor to repair the damages.

Sometimes employers add items to the Snag List, or Punch List, which are additional to the contractor's scope. It's important the contractor's team confirms that all items on the Snag List are due to their fault or omission. If there are additional items the contractor needs to notify the employer that the item is extra to the Contract Scope and a Variation Order must be issued. Failure to obtain an Instruction or Variation Order to attend to these items may result in the work going un-paid.

Claim mitigation

Frequently Contract Documents state that contractors must take mitigating actions to avoid delays. Mitigating actions aren't usually defined and some employers interpret this to mean that contractors shouldn't allow work to fall behind schedule. In essence, this could mean that even if the employer has delayed the project, or the contractor has been delayed due to causes beyond their control, they should accelerate the works to avoid the delay. In other words the contractor should catch up the delay at their cost. This is obviously nonsense and if it was so we would never be able to submit Extension of Time Claims.

However, the mitigation clause does place an onus on contractors to take action to avoid delays. This could include providing the employer with an Information Required Schedule and reminding the employer well ahead of time that access or information will be required shortly, therefore giving the employer time to rectify the situation, thus avoiding any possible delay. It's important that these notifications are in writing and are included in site meeting minutes.

Subcontractors

Frequently subcontractors lodge Claims against the contractor. Important documents that should be kept to mitigate and refute these Claims include:
1. Instructions to subcontractors.

2. Drawing distribution records and receipts.
3. Daily Diaries from the subcontractor which record:
    a. Manning.
    b. Equipment.
    c. Progress.
    d. Delays.
4. Minutes of regular meetings held with subcontractors which should record:
    a. Progress measured against the subcontractor's approved Construction Schedule.
    b. Delays – it's important that if there were no delays that this is recorded as such.
    c. Quality control and test results, including any quality problems.
    d. Safety, including any incidents and problems.
    e. Outstanding information – record this as none if there is no outstanding information.
    f. Variation Claims from the subcontractor.
5. Signed acknowledgement of handover and acceptance of the area or section of work to the subcontractor.
6. Letters and emails sent to the subcontractor and received from them.
7. Receipts of material issued by the contractor to the subcontractor.
8. Non-conformance reports for the subcontractor's defective work.
9. Dated photographs of the subcontractor's work

The subcontractor should submit a Construction Schedule for their portion of work. This Schedule should be checked by the contractor to ensure that it:
1. Complies with the overall Construction Schedule for the project.
2. It doesn't require information from the employer, or contractor, to be issued earlier than specified in the subcontractor's Contract Document, unless this has been agreed.
3. It has allowed for discontinuities and project restraints included in the subcontractor's Contract Document.
4. It allows for interfacing with the contractor's work as well as with other subcontractors.
5. It allows for approval of the subcontractor's design and drawings where this is required. This approval time shouldn't be shorter than that stipulated in the Contract Document.
6. It's logical and achievable.
7. It allows for all the specified quality checks and tests.

8. It's allowed for commissioning of the subcontractor's work and any tie-ins with existing and new services and facilities.

It should be noted that in some cases subcontractors have to manufacture items off the project site, or have to procure Long Lead items. The Schedule should include these items so that their progress can be monitored.

All instructions to the subcontractor must be in writing and should be issued to the subcontractor's designated responsible person. Quality problems should be recorded on non-conformance reports and the subcontractor must be advised immediately of quality problems so they can take corrective action and prevent the problem from reoccurring.

Services, facilities, materials, equipment and other resources which the contractor has supplied to the subcontractor, which the subcontractor is responsible for and should have supplied, must be recorded and agreed with the subcontractor on a daily basis. These must be charged to the subcontractor on a monthly basis. These records should record the date, the item, hours worked and the rate for the item.

Insurance

The contractor needs to know what insurances the employer has in place to cover the construction works. This includes, understanding the terms of the policy, what's covered and what's excluded, the duration of the policy, and what the excesses (deductibles) are on the policy.

The contractor should evaluate the risks and check that the appropriate insurance cover is in place. It may be necessary to arrange additional insurance should the employer's insurance be inadequate, or if the policy excesses or deductibles are too high.

There are various insurances a project should always have, these include, amongst others:
1. Equipment insurance.
2. Insurance of the works.
3. Public liability insurance.
4. Workers compensation insurance.
5. Professional indemnity insurance if there's design involved.

The contractor's existing policies must be reviewed to ensure they adequately address the risks on the project. If there are any doubts as to what insurance cover should be in place the contractor should check with qualified experts on the matter, such as with the contractor's insurance broker.

Claims could be repudiated by the insurer if all mitigating steps weren't taken to prepare for, and prevent an insured event from occurring. For instance, if a vehicle is in an accident and it was found the driver didn't have a valid licence, the vehicle wasn't properly maintained or was un-roadworthy the insurer probably won't pay for damages. The same applies to items damaged in a flood where the contractor didn't take proper precautions to prevent the works from being flooded, or, didn't move equipment ahead of a forecast flood.

Equally important is to ensure that the insurer is given accurate information when the insurance policy is purchased and that they are advised of any changes in conditions that occur in the course of the project. This could include Variations, Extension of Time, Scope change and changes in the methods of construction. Failure to do this could render the policy null and void. Of course it's necessary to ensure that the premiums have been paid.

It's also important to check that all hired equipment is adequately insured.

Archiving records

At the end of the project all documents should be archived by the contractor where they can be easily accessed in the event a Claim arises, and in a secure area where they can't be damaged. Unfortunately Claims often arise months, or even years after a project has been completed. Valuable documents that should be kept include:

1. A full set of construction drawings including all previous revisions.
2. The 'As Built Drawings'.
3. All correspondence between the contractor and the employer, and between the contractor and subcontractors.
4. The Contract Document.
5. The Construction Schedule.
6. Updates of the Construction Schedule.
7. Minutes of meetings with the employer and with subcontractors.
8. All test results.
9. Daily Diaries.
10. Project Reports.
11. Diaries kept by the Project Manager and Supervisors.
12. Invoices from suppliers and subcontractors.
13. Records of payments to suppliers and subcontractors.
14. Weekly payrolls or records used to calculate the payment of workers.
15. Equipment time sheets or records used to calculate the payment of equipment. Including records of equipment breakdowns.
16. All Variation Claims including supporting documents and calculations.

Chapter 2 – Supporting Documentation | 65

Summary

The contractor needs to ensure that documents are gathered from the start of the project, that these are accurate and that they are stored where they are safe and can be easily accessed by the contractor. These documents include:

1. The Contract Document.
2. Tender information.
3. The Construction Schedule.
4. Instructions.
5. Letters.
6. Emails.
7. Daily Diaries or Logs.
8. Personal diaries.
9. Request for Information.
10. Weather records.
11. Meeting minutes.
12. Photographs.
13. Project reports.
14. Records of Variation work.
15. Hand-over documentation.
16. Drawing registers.
17. Construction drawings.

Accurate documentation plays a significant role in supporting the contractor's Claims, or as evidence required to refute Claims lodged against the contractor by the employer or subcontractors.

Chapter 3 – Preparing the Claim

Generally contractors have only one opportunity to convince their employer that they have a valid Variation Claim, and then to prove the costs associated with their Claim. Once the employer dismisses a Claim it's difficult to convince them that they made the incorrect decision. Minds are often made up and thought processes closed. There may even be 'face to lose' by admitting they were wrong to reject the contractor's Claim. In some instances they've already informed their management the Claim is invalid and aren't happy to tell them they've reversed their decision and approved the Claim, which could be perceived as they're weak or aren't competent.

It's therefore essential that the Variation Claim is carefully thought out and presented clearly with all the relevant supporting documents. The value should be substantiated and should include all of the relevant costs. Once the employer has approved the Variation Claim it's near impossible to go back and ask for additional costs that the contractor forgot to add into their original Claim.

Of course it's important that the Claims submission process complies with the Contract Document. Failure to comply with the Document, or lodging the Claim incorrectly, may result in it being rejected on technical grounds.

Generally the contractor needs to demonstrate in their Claim that:
1. The work performed was different to the Scope in the Contract Document.
2. The employer was notified of the change.
3. The change was required by the employer and not volunteered by the contractor.
4. The contractor put in place mitigating measures.
5. There were actual costs or time impacts to the contractor, including what these were, and how they were formulated.

The Variation Claim process

Often employers request the contractor to submit a Variation for additional work. This request could come in the form of an Instruction, Change Order Request or Request for Change Proposal. But more often the contractor detects a Variation when a delay event occurs, or a new drawing is issued with information different to what was originally priced, or the contractor notices different Contract or site conditions.

The onus is on the contractor to immediately notify the employer that there is a Variation (see below notifications). The Claim notification often doesn't include the quantum of the Claim (the costs or the time impact) unless these are immediately evident to the contractor. Rather the notification would make reference to the Variation Claim arising and that there will be time or cost impacts which will be issued to the employer when the contractor has calculated their quantum, or within the time period indicated in the Contract Document. The notification serves as an Early Warning to the employer and provides them with an opportunity to withdraw or change their Instruction, drawings or specifications.

Once notification has been lodged the contractor has to quantify the magnitude of the event. If there's a time element as well as a cost element which varies according to the amount of time granted, then, it may be relevant to first submit the Extension of Time Claim and agree that, before submitting the costs associated with the agreed Extension of Time. When submitting a time Claim separate from the associated cost Claim it's important for the contractor to note on the Claim that the costs will be submitted separately once the time portion is resolved.

If the contractor is unable to quantify the time or costs within the time specified in the Contract Document they need to apply to the employer for an extension for the lodgement of the Variation Claim. Sometimes, the contractor is only able to submit an estimate of the costs or time of the impacts, but is unable to finalise the Claim because of missing information. These estimates must be clearly marked as an 'estimate', noting that the final impacts will be submitted when they become known.

Once the employer and contractor agree on the cost and time impacts the employer normally issues a Variation Order, Change Order, or Contract Amendment. This is required so the contractor can be paid for the Variation. This is effectively a modification to the Contract Document. The Change Order is only valid if it's signed by the employer's authorised person and signed and accepted by the contractor. It should be noted that the value of the Change Order may dictate who from the employer is authorised to sign, because some representatives might only be allowed to sign Change Orders if they are below a specified value. A higher authority may have to sign a Change Order with a higher value. (It should be noted that on some occasions the employer's representative could try and circumvent this ruling by issuing two Change Orders whose total value equates to the agreed value of the Variation but whose individual values are less than the person's limit of authority. Contractors should take care agreeing to this, particularly if it's clear this method was chosen to circumvent the limits of authority. An audit by the employer may nullify any value that exceeds the person's limit of authority.)

Preferably the Change Order should have the agreed value of the Change Order as well as the new revised Contract Value. The revised Contract Value is the sum of the original Contract Value and all the previously agreed Change Orders as well as the new Change Order value. It is important the contractor checks that this value is correct because this will be the value the contractor is paid once they have completed all the work on the project.

Variation Claim Notification

The employer must be notified of Variations as soon as the contractor becomes aware of them, and certainly within the time specified in the Contract. Failure to do so may mean the contractor loses their right to lodge a Claim.

The Claim must:
1. Be lodged within the time-frame specified in the contract.
2. Clearly state the reason for the Claim.
3. Be addressed to the correct person.
4. Be delivered to the correct address.

It's often good practice to discuss large or potentially contentious Claims with the employer before submitting them. This not only forewarns the employer that the Claim is coming, but also provides an opportunity to discuss the reasons and the merits of the Variation Claim.

In some cases the employer may try and deter the contractor from lodging the Claim, but contractors shouldn't be dissuaded if they believe the Claim has merit. It may not be possible to lodge the Claim later, but it's always possible to withdraw a contentious Claim at any stage. I have had employer's representatives yell at me because they were angered by a Claim, yet, once presented with all the evidence they understood that in terms of the Contract our Claim was valid and it was later successfully agreed with them.

Time-Bar

Most Contract Documents specify the time within which a Variation Claim must be lodged. This is normally expressed as the number of days after which the contractor became aware of the event. It's advisable that contractors comply with this time period. However, it should be noted, that exactly when the contractor became aware of an event is open to interpretation.

> For example: *the employer issued the contractor with a revised construction drawing that increased the size of a structure which gave reason for the contractor to claim an Extension of Time. Does the time start when the contractor received the drawing or when they actually opened the drawing and noticed the change? The wording is almost always when the contractor became aware of the event so by interpretation it's when they noticed the change and not when the drawing was issued.*

It's possible for the contractor to successfully fight a Time-Bar to their Variation Claim, particularly if the employer has been 'enriched' by the event. However employers can equally successfully argue that since the contractor didn't comply with the notice provisions they were unable to put mitigating measures in place, or take avoiding actions, even reversing the event, which would have limited or even prevented the Variation Claim from arising.

Another difficulty with Time-Bar provisions is who from the contractor should become aware of the Variation event. If the contractor's Supervisor notices an obstacle restricting access to a work area is that when the notice period starts? Convention is that it's when the contractor's managers or directors become aware of the claimable event. This would generally be interpreted as the contractor's Project Manager or Site Manager.

To avoid arguments the contractor's manager should ensure they are aware of all changes, Variations and delays on the project and that the employer is notified immediately of these. The contractor's team must also be aware that they should notify their Project Manager immediately they notice a delay or a possible Variation.

Put the effort in

I've seen many poorly put together Variation Claims – some for millions of dollars. Yet construction companies seem to think that they can apply minimal time and effort to formulating their Claims. That it will be easy money! The employer will agree their Claim – they just have to! No they don't.

Some contractors even delegate their Claim preparation to junior inexperienced Quantity Surveyors or Contract Administrators. Sometimes the contractor's Project Manager isn't even aware a Claim has been submitted and is blindsided when their employer takes offence to a spurious or an unsupported Claim.

Some of these Claims are poorly thought out and presented. I've seen Claims with arithmetic errors, spelling mistakes (even incorrectly spelling of the employer's name), factual errors, contradictory information, confusing language and unsupported evidence.

Consider how hard a contractor has to work to earn a million dollars, or even ten thousand dollars, so why then do some construction companies expect to spend minutes, or an hour, preparing a Claim for the same amount.

Construction companies should ensure that a knowledgeable and experienced person is allocated to draft the Claim (one familiar with the Contract Document, the employer, and the work that's been constructed) and that it is reviewed and checked by the contractor's Project Manager before it's submitted.

If there's any doubt as to the validity of the Claim, or what should be included, seek advice from experts within the company or from outside providers. The cost of getting proper advice is often far outweighed by the revenue that can be earned by an expertly formulated and drafted Claim.

What should be included in the Claim

Generally Claims should be kept simple, with one event covered by each Claim, unless the events are linked. A properly drafted and well thought out Claim will be hard to refute and it's likely to be successful.

To increase the chances of a successful Claim here are some points to consider. Claims should have:

1. A description of the event.
2. The cause of the event.
3. The date of the event, where relevant.
4. Clauses in the Contract Document relevant to the event.
5. The full impacts of the event – it may be necessary to detail all the impacts because employers often don't understand how one small change can sometimes have a ripple impact across many tasks which magnify the ultimate total impact.
6. Steps taken to mitigate the impact. This could include letters or meeting minutes which refer to the risk of the event occurring.
7. The cost and time impacts of the event.
8. All the supporting documentation attached (or should refer to supporting documentation correctly referencing the relevant Contract clause numbers, construction drawing numbers, Schedule or programme task numbers, correspondence, Bill of Quantity items or Contract specifications as required).

It's essential that this supporting documentation is relevant to the Claim, supports the Claim and is not contradictory (any contradictions must be explained as part of the Claim).

As part of formulating the impact of the event all calculations and Schedules should be included. The Claim Schedule should reference the approved Construction Schedule. Calculations should refer to where the facts and figures came from, how they were derived and how they were put together. The calculations should be checked for arithmetical errors (which can occur all too frequently).

Claims should be written clearly in the language of the Contract. It's usually not necessary to add legal, or pseudo legal, terminology which often creates confusion and may even irritate the person assessing the Claim. Keep all arguments short and concise.

It's good practice to ensure that the scope of the Claim is clearly outlined. Sometimes there are a number of closely related or overlapping Claims and the employer may become confused and approve one Claim thinking it included Claims for other delays and Variations on the project.

The letter should refer to previous correspondence relating to the Claim, including previous notifications submitted.

It's important for contractors to remember that the person adjudicating the Claim might not be familiar with the project specifics, or the events that gave rise to the Claim. Therefore contractors should ensure that the Claim clearly outlines the events leading to the Claim as well as the full impacts of the Claim. Should the Claim end in a Dispute it could go to arbitration or other processes where impartial outsiders will investigate the Claim. If they don't understand the Claim, or how time delays or costs were arrived at they could reject the Claim. Don't assume the person reading the Claim has prior knowledge of the facts that resulted in the Claim, or is familiar with events as they occurred on the project.

When preparing a complex and lengthy Claim it is useful to include an index. Pages should be numbered, and clauses and paragraph numbering must be consistent and sequential. Presenting the Claim in a folder or in a bound document not only looks professional but it also ensures that pages don't become muddled or lost. When presenting the Claim in an electronic format ensure the employer doesn't have access to documents and worksheets that they shouldn't receive.

Don't assume an earlier Claim will be approved

Frequently there are many Claims on a project, some of which overlap. Some of these Claims could take a long time for the employer to respond and for

Chapter 3 – Preparing the Claim

agreement to be reached. When formulating Claims the contractor doesn't always know if previous Claims will be accepted. If they overlap does the contractor only base the new Claim on the portion that extends beyond the first Claim or do they assume the first Claim doesn't exist or will be rejected.

> *For example: in the month of May the project experienced a 5 day delay due to inclement weather for which the contractor submits an Extension of Time Claim for 5 days. This Claim hasn't been agreed when in June the employer supplies information 9 days late according to the agreed Construction Schedule. If the Claim for inclement weather had been agreed then the information would only be 4 days late according to the revised Construction Schedule adjusted for the weather delay of 5 days. However, since the weather delay Claim hasn't been agreed the contractor must submit an Extension of Time Claim for the full 9 days the information was late. They cannot assume that the weather related Extension of Time Claim will be accepted and only claim 4 days, because if the weather Claim is rejected they won't receive the full Extension of Time they are entitled to. This can become confusing because the contractor may have several Claims waiting to be agreed. In this example it appears as if the contractor has one Claim for 5 days plus another for 9 days which seems the total Extension of Time requested is 14 days. However, since the Claims overlap the total is only 9 days. To overcome any confusions and misconceptions the contractor could add a paragraph to their 2nd and subsequent Claims summarising the impact of previous Claims together with the current Claim. This summary could say that the impact of this Claim on its own is 9 days, but if the previous Claim (insert the date and number of the Claim) for 5 days is agreed then this Extension of Time Claim would be reduced to 4 days and the overall Extension of Time Claim for the two Claims combined would only be 9 days.*

Until a Claim is agreed in writing the contractor has to submit all the following Claims assuming the Claim has not been accepted.

Extension of Time Claims

All Extension of Time Claims should be presented using the approved Construction Schedule. If the activities on the Schedule are correctly linked and sequenced and have the correct durations, demonstrating a case for an Extension of Time can be simple. If the Schedule is correctly resourced, and the employer has access to this data it should also be simple to demonstrate a case for Acceleration and the mobilising of additional resources should these be required.

The Variations may be due to the contractor receiving drawings, information or access late, weather events or unforeseen interruptions on the project, or the project Scope changing as a result of additional activities or an increase in quantities. The Claim must be due to the employer, their representatives or unforeseen claimable events which the contractor wasn't responsible for.

All delays should be individually entered onto the Schedule which should then automatically show the effect of the delay on the project duration. Some delays may be Concurrent and these may have to be entered onto the Schedule together so that any concurrency is taken into consideration.

Most Contracts and employers will only accept delays to the Critical Path and those that affect the overall contract duration. It's important that the effect a delay has on the Critical Path is clearly demonstrated. Also as discussed previously, most employers want to see the Schedule Float consumed before they'll accept that there's been a project delay.

There are a number of ways of entering a delay on the Construction Schedule.

1. When information or access is granted late it could be entered as a milestone date ahead of the activity it immediately impacts. With information there's often a Lead Time for procurement which can either be entered as another milestone or as an activity with duration equal to the Lead Time. Remembering that when a construction drawing is issued it could impact several tasks and there could be different Lead Times for the different tasks.

    For example: a concrete drawing could include the dimensions to the structure which say could have a 5 day Lead Time for the start of the formwork installation. The Reinforcing may have a 10 day Lead Time to procure the reinforcing which would mean installing the reinforcing could only begin 10 days after receipt of the drawing. The drawing may include cast-in steelwork items that may have a 15 day Lead Time meaning that the structure could not be concreted until these items have been procured and installed.

2. A weather delay could be entered as a non-work day on the Schedule calendar. The problem with this is it will impact all activities even those that weren't actually impacted by the weather, such as those happening off-site or undercover.

3. The alternative is to add the impact of the weather delay to every activity impacted by it and extend the activities duration.

    For example: if work was delayed by 1 day then the duration would be increased by 1 day.

Unfortunately adding additional days to activity durations isn't always obvious and creates confusion.
4. Adding in additional activities. This is sometimes required when a Variation entails additional work or completed work has to be redone to match a new drawing. It's sometimes possible to add in the impact of a weather delay as an extra activity but this can become messy on a project with multiple weather delays and thousands of tasks.

It is usually not wise to change or add links between activities as these changes are often not obvious and it can create confusion.

Delays not on the Critical Path should be shown because when future delays are added they could ultimately impact the Critical Path and the Completion Date.

Factors to consider when formulating an Extension of Time Claim

Most contractors believe that using the approved Construction Schedule, with linked tasks and a Critical Path, they can simply input the delays and automatically get a new end date. Essentially this is true but it doesn't take into account some factors such as:
1. The revised Schedule may push certain activities into unsuitable seasons.
   For example 1: the contractor may have planned to complete the excavations and earthworks in the dry season but because of delays these activities are now pushed into the wet season. In some areas it can be impossible to carry out earthworks in the wet season, but at the very least the rains will cause additional delays which weren't allowed in the original Construction Schedule.
   Example 2: the original Schedule was prepared on the premise that the building being constructed would be weather tight by the time severe cold or rains arrived. Delays to construction could result in the building not being weather-tight before the onset of poor weather resulting in weather related delays which weren't allowed in the Construction Schedule.
2. The revised Schedule can cause discontinuities and equipment such as large cranes may have to be re-established again or could be under-utilised for a period.
3. The revised Schedule may have resource peaks and troughs. This might result in resources (men and equipment) having to be removed and re-established on the project resulting in additional costs.

4. Pushing out the end date of the project may mean that specialist subcontractors and equipment might not be available as they are already committed to other projects.

   Case study: On one earthworks project that was both delayed and increased in Scope, most of the equipment had been ordered for the original project duration and was committed by the supplier to another project at the end of this time. At the end of the original contract period the supplier removed all their equipment, even though we weren't finished using it, and we had to source new equipment which caused us delays and additional costs.

5. Materials that have already been ordered may have to be delivered to the project site in accordance with the original procurement order, even though they are no longer needed at that time. This could result in the site becoming congested. Also, the materials probably now have to be double handled, possibly necessitating additional transport and cranes to move them. The alternative is to store the materials and equipment off-site at an additional cost which would include security and unloading and reloading them. There's also the additional risk of damage to these items with the extra handling and storage. In addition the contractor usually has to pay for these items because they're considered delivered by the supplier, but the employer may not pay the contractor for them until they are installed, which will be in accordance with the delayed Construction Schedule – with sometimes dramatic consequences for the contractor's cash flow.

6. Moving the Construction Schedule out could also move the Schedule to coincide with statutory public holidays. These holidays cause further delays. In some countries projects shut-down for an extended period over Christmas. Again this period further delays the project. But in addition if the original plan was to have finished the work before this holiday shut-down the contractor will now face additional costs of having to secure the project site and maintain their site facilities over this period.

7. In some cases the contractor may have planned their Schedule so that major concrete pours occurred before the holiday period. The concrete was planned to cure and gain strength over the break so no time was allowed in the Schedule for this activity. A delay may thwart this plan meaning that the concrete is only poured after the break and the curing time has to be added to the Schedule. A delay of a couple of days could impact the schedule by several weeks.

Chapter 3 – Preparing the Claim

What is a reasonable duration for additional tasks?

Case study: I recently visited a new office building under construction which had a void 5 floors high over the entrance lobby. The contractor had constructed the concrete roof slab spanning the void which was temporarily supported from the ground floor. The concrete had achieved its design strength and the contractor was about to remove the support-work to this slab when concerns were raised regarding the deflection of the newly completed concrete roof. The employer's engineer elected to add in 2 additional structural brickwork walls to support the roof. These started from the ground and went to the 5th floor. The support-work for the slab over the lobby had to stay in place until the columns had sufficient strength. This obviously had severe time implications for the lobby area which was on the project's Critical Path. In submitting their Extension of Time Claim the contractor had to make the following decisions:

1. *How much time should they claim to build each wall?*
2. *Could they build the 2 walls simultaneously?*

The agreement of Extension of Time Claims frequently occurs after the claimable event has been completed. In this case, for instance, the contractor could have claimed 15 days to construct the walls. If they actually took 12 days then when the employer came to assess the Claim (which was after the walls had been constructed) they would probably only grant an Extension of Time of 12 days and not the 15 days the contractor had claimed. Unfortunately the converse doesn't usually happen and if the contractor claimed 15 days but actually took 18 days to complete the work the employer would probably only agree to the 15 days originally claimed by the contractor.

In this example it would seem that it's better for the contractor to claim a longer duration for the additional activities knowing that the Extension of Time would be agreed after the additional work has been completed, and assuming that the employer would eventually agree to an Extension of Time that equated to the actual time taken to complete the additional tasks. However, it should be noted that the employer does not have to accept the actual time the contractor took to complete the task as being the time that they should approve for the Extension of Time. The employer could successfully argue that the contractor was inefficient or could have used alternate methods that could have shortened the duration and reduced the Extension of Time claimed. So in this case even if the contractor took 15 days to build the walls the employer may only agree to

> an Extension of Time of 12 days if they could prove the work could have been completed in this shorter time.

Indeed, when dealing with most delays and Extension of Time Claims, time is of the essence and the contractor needs to complete the additional work as quickly as possible. However, this does put the contractor in a quandary as how quickly is quickly and what costs will the employer accept as being reasonable to achieve the completion of the activities in the shortest possible time?

> *Case study continued: so with the previous case study the contractor has to decide can they construct both walls at the same time? To do this may require them to bring additional resources to the project which entails additional mobilisation costs. In this case deciding what is best for the project may be as simple as comparing what it will cost the employer in additional mobilisation costs compared to the cost of the Extension of Time. If the cost of the additional resources is less than the delay costs then the answer is usually simple and it makes sense to bring the additional resources to the project and add these costs to the Claim.*
>
> *However, it isn't always as simple as this! In this example at some stage the lifting of the bricks and the mortar to the elevated platforms for the construction of the brick walls can become a limiting factor. On this project the only crane on the project was now allocated to installing the external cladding to the building. Using the crane to lift bricks and mortar for the additional walls now disrupts and delays the external cladding installation. Does the contractor bring additional lifting equipment to the project? What costs will the employer accept to limit the delay to their project? Certainly the contractor could look at all kinds of other alternatives to reduce the duration of constructing the additional walls which might include using quick drying mortars, double shifts or even introducing extra propping systems. At some stage these additional costs will far outweigh any cost advantage the employer gets from the reduction of the delay.*

When deciding how long the additional tasks will take to complete contractors have to:

1. Consider the available resources and impacts on the rest of the project.
2. Use appropriate construction methods which will meet the quality and safety standards.
3. Consider alternative solutions which may be shorter.
4. Use durations which are reasonable and which can be justified.
5. Carefully weigh up what is more important to the employer – minimising the costs or minimising the delay. Often a compromise solution is the best option.

To achieve the best solution requires an understanding of the employer's requirements and restraints. It often pays to discuss the alternatives with the employer, but bearing in mind decisions should be made quickly so that the delay isn't exacerbated waiting for the employer's reply. It's good practise to confirm these discussions in writing so that when the Claim is assessed the employer can't argue that the costs claimed to reduce the delay were excessive.

The spatial arrangement of structures and delay impacts

Contractors and employers often don't consider the impact of changes to one structure on the adjacent structures. Often these structures aren't linked because when the Construction Schedule was prepared the structures had no impact on each other. However, a change to one structure could cause it to now impact an adjacent structure.

Example: Consider a building which consists of a number of foundations in close proximity, all founded at the same depth below ground level. The approved Construction Schedule could have had the foundations being constructed as one task at the same time. If the employer issued new information where one foundation is now lowered a metre below the others we may assume that the only extra time required is to excavate the foundation a metre lower and we could extend the task duration accordingly. However, if the deeper excavation now means the other foundations can't be constructed until the deeper one is completed and the foundation is backfilled with soil to the underside of the other foundations we now have to introduce new activities and new links. Clearly the delay will be more than just excavating a deeper excavation, but will also include the time taken to complete the deeper foundation, before the other foundations can begin.

Changing one part of a structure, or changing an adjacent structure, or delaying one structure or changing the order in which structures are planned to be constructed can have profound impacts on adjacent structures, causing delays beyond the structure that was first changed or delayed. It's therefore important to understand these impacts fully.

Changes and their impacts on access

Often contractors don't consider how changes can impact access to the work area and consequently delay the Schedule resulting in cause for an Extension of Time Claim.

Chapter 3 – Preparing the Claim | 79

Example 1: the employer changes a building's roof from a steel structure to a concrete slab. For the purposes of this example let's assume that the steel roof and the concrete roof take the same amount of time to construct. Initially there appears to be no delay. But, a concrete slab needs to be supported for a period of 7 to 10 days while the concrete gains sufficient strength. While the support props are in place work probably can't continue under the roof slab. This is additional time which has to be added to the Construction Schedule.

Example 2: The employer adds an underground sewer line to the contractor's scope which they want installed as soon as possible. At first glance the contractor could easily add this anywhere in the Construction Schedule with no impact to the completion date. However, the excavation for the sewer could disrupt access to the contractor's other work areas preventing materials from reaching the areas or blocking access for cranes and equipment.

It's important to consider the full impacts of any delay and Variation on all the tasks on the Construction Schedule.

Deciding on which Claim to focus on

Sometimes a number of delay Claims can occur almost at the same time and run concurrently. It's important for the contractor to understand which Claims will be most favourable and under which clauses they should be pursued.

For example: the employer has delayed the project by issuing information late, but a few days before that the project was impacted by bad weather, or a Force Majeure event, which overlaps the late information. The contractor may choose to ignore the weather or Force Majeure delays because they might only be granted an Extension of Time for these delays without costs. Whereas the employer's late information will result in an Extension of Time Claim with costs being accepted. (I discussed concurrency and precedence of delays in Chapter 1 and mentioned that concurrent Claims are normally considered in the order they occur.) Considering the weather and Force Majeure Claims first may lessen the impact of the late information delay which came afterwards.

There's a risk to the contractor if they haven't claimed for the other delays (such as the weather event in the above example) and the employer rejects their late information Claim. It may be too late then to submit a Claim for the other delay event. Of course the employer may introduce the weather related Claim even though the contractor hasn't, but I've seldom found employers first raise an Extension of Time Claim.

Acceleration Claims

Before agreeing to an Acceleration Schedule contractors need to ensure that the new Schedule is achievable. Frequently contractors are bullied into accepting new dates by the employer which might not be achievable. This is dangerous since once the Acceleration Schedule is agreed Liquidated Damages could be applied if the contractor doesn't achieve the new completion date. Another potential problem is the approved acceleration cost is based on the contractor completing the project on the new designated date. Failure to achieve the date may mean the contractor isn't entitled to payment for Acceleration even if the work has been completed ahead of when it would have been completed without Acceleration.

> For example: if there is an agreement to pay the contractor fifty thousand dollars to Accelerate the project by 10 days, but they fail to achieve this and complete the project 2 days later (in effect achieving an Acceleration of 8 days), their claim of fifty thousand dollars for Acceleration could be refuted and the employer could impose Liquidated Damages.

To prove how much the project is being accelerated it's advisable to include the 'un-accelerated Schedule', showing when the project would be completed with the delays and new Scope, working at the same rate of production as the approved Contract Schedule. The employer will then see how many days the Acceleration Schedule will save them.

Putting together and proving the costs of acceleration is often difficult. Acceleration can be achieved by doing a number of different things including:

1. Working longer shifts.
2. Working 2 or more shifts in a day.
3. Working on non-work days such as weekends and statutory holidays.
4. Increasing the size of the crews.
5. Using more equipment.
6. Increasing the size of equipment – for example using larger excavators and trucks.
7. Fabricating portions of the project in modules either on, or off-site.
8. Changing the construction methodology – for example making use of precast components.
9. Overlapping trades (stacking trades), meaning more trades may have to work in the same area.

10. Making use of different materials – for example using rapid hardening cement or higher strength concrete to enable the earlier stripping of formwork.

Costs to accelerate could include:
1. Mobilisation costs for additional equipment and people, including; transport, inductions and non-productive time.
2. Ongoing costs for the additional employees that might also include their accommodation and daily transport.
3. Overtime costs (penalty rates) for employees working longer shifts, additional shifts or on statutory holidays.
4. Loss of productivity caused by:
    a. Congested work areas.
    b. Employees working longer shifts. (I've generally found that production from tradespeople and general workers decreases after a nine hour shift.)
    c. Working at night.
    d. Workers and equipment not having continuity, or waiting for other trades to finish.

    This loss of productivity is often difficult to quantify and prove.
5. The additional costs to change the methods of construction which might include design changes, additional transport, more expensive materials and specialist equipment.
6. Subcontractor and supplier costs which they incur to accelerate their portions of work.
7. Lighting to work areas at night.
8. Additional supervision and management to facilitate the acceleration.
9. Bonuses and incentives to improve production, or to get employees to work additional shifts.

It should be noted that in some cases acceleration may in fact improve productivity as the equipment could complete more work per shift than if the work wasn't accelerated.

Costs for Extension of Time Claims (prolongation costs)

There are a number of costs to consider when formulating an Extension of Time Claim which includes:
1. The contractor remaining on the project longer than allowed, thus incurring additional costs for:
    a. Their site facilities such as offices, stores and ablution facilities.

b. Staff, such as management, supervision and administration personnel. These costs could include; their base salary, plus entitlements (including leave pay, sick leave, insurance and bonus provisions), and vehicles and accommodation (where provided).
   c. The extension of the period for bonds, sureties and insurances.
   d. General wage costs which aren't recovered in the prices for the individual construction tasks. These could include; storemen, drivers, cleaners and crane operators. These costs would include the hourly rate plus entitlements (including; overtime, leave pay, sick leave, insurances, bonus entitlement and special allowances), and accommodation and transport where provided.
   e. General construction equipment costs which aren't recovered in the prices for the individual tasks, such as; cranes, hoists, site vehicles and generators. If these items are parked-up they normally wouldn't include fuel costs and wear and tear costs. However they could include insurance costs.
   f. Major temporary works which aren't recoverable elsewhere such as; dewatering, access towers and scaffolding.
   g. Site services such as the cost of water and electricity.
   h. The costs of security, safety and traffic control if these are ongoing costs during the delay.
2. Offsite overhead costs which consist of two distinct components
   a. Those directly associated with the project which might be Head Office based project staff such as; Administrators and Quantity Surveyors which are split between a few projects, including the one that has been extended. It's often difficult to prove to employers who the Head Office based project employees are, or why they couldn't be paid in full by one of their other projects. Of course if it's possible to use these people on other projects in the interim it may not be necessary to charge the employer for them. If the contractor isn't charging for them they should still be shown in the calculations but as a nil cost. This demonstrates good will on the contractor's part, but, also in future delays the contractor may not be able to utilise these people elsewhere and will have to charge for them, which could raise a query for the employer if they weren't shown on the previous Claim.
   b. Indirect overheads which are the costs of running the company and Head Office. Most contractors apportion these costs to each of their current projects. These costs are normally claimed as a

percentage and should probably be incorporated in the contractor's profits and overheads added to the total Extension of Time costs.
3. The inefficient and unproductive use of personnel and equipment which:
   a. Cannot be used at all because of the delay.
   b. Are only partly utilised.
4. If the work moves into a season with unfavourable weather conditions which wasn't allowed for in the Schedule or Tender, then the costs and delays associated with these poor weather conditions.
5. Material, equipment and wage costs increasing in the interim. (Obviously contractors can only claim for these increased costs for the portion of work that would have happened before the increases became effective and which now can only occur after the implementation of the increases.)
6. Productivity losses caused by activities being delayed so they now happen when other contractors are working in the area.
7. An activity being undertaken out of sequence which may result in:
   a. Access being limited from what it would have been if it was completed as per the original Schedule. This could either reduce the productivity of the activity, possibly adding to the time and the cost to do the work. It could even mean alternative methods, or different equipment, has to be used, which might be more expensive, or, it may require other equipment to be mobilised to the project with the associated additional costs.
   b. The work area becoming congested due to other activities happening simultaneously which impacts productivity possibly adding to costs as well as the time to complete the task.
   c. Specialist equipment, subcontractors or personnel not having continuity of work resulting in them having to return to the site at a later date entailing additional mobilisation costs. In some cases, the equipment or subcontractor might not be available when required again, which could result in further delays.
   d. Damage to works already completed caused when the delayed activities are finally constructed.
   e. Additional protective measures to protect completed work.
8. Materials which have been ordered having to be stored because the site isn't ready to use them, resulting in storage costs and double handling, and the associated risks of damage and theft of these materials. This could include protecting the items from the weather.

9. Disruption of cash flow because the project's end date is extended, deferring the release of retentions and securities. A simple calculation would calculate the additional interest costs of the retention monies for the duration of the Extension of Time.
10. Subcontractors are delayed resulting in them claiming their delay costs.
11. The contractor's profit and overheads on the costs above.

Some of these costs may be difficult to demonstrate and prove to the employer. Frequently, employers don't understand the consequences of their actions. However, keeping good daily records of personnel and equipment on the project is helpful. Contractors should ensure that the basis of these costs is clearly explained in their Claim.

It should be noted that in general the contractor cannot claim for the cost and time of preparing the Variation Claim and nor can the employer claim back monies for reviewing the Claim, even if they can prove the Claim is unfounded.

In some cases contractors can claim for loss of profit. Although in many cases there is likely to be no loss of profit as the contractor would in most cases be paid the original Contract Value plus more for the Extension of Time costs. However, the contractor may claim loss of opportunity as it could be argued that the resources that have to remain longer on the project could have been used elsewhere on another project. This is generally difficult to prove and the contractor would have to demonstrate that they turned another project away, that they were certain of being awarded, because their resources were still tied up with the delayed work. Of course if the Extension of Time was due to the project growing in Scope then the contractor would usually have no cause to claim for loss of profit as the Contract Value would increase in size delivering a larger profit and overhead than the original Contract Value.

Contractors should tread carefully so they are seen to fairly claim their justified actual costs without claiming for every trivial item.

Mitigating losses

Just because an item is delayed on the project doesn't automatically entitle the contractor to be compensated for the delay. Employers will expect the contractor to take mitigating actions to reduce the costs incurred while the project, or section of works, is delayed. This is sometimes difficult for contractors as they usually can't just send their employees home without pay, or put equipment off-hire, and then hope that it's available when required. Employers often think that contractors have a rack of people and equipment which they can draw from as required and return when not required. This obviously isn't the case. What makes this more awkward is that sometimes the delay is indefinite and the contractor doesn't know when work will be able to resume again on the project.

Both contractors and employers have to be reasonable when it comes to mitigating losses. This reasonableness includes the contractor sending people and equipment off-site when possible. It also includes discussing with the employer where the employer could utilise these resources more effectively.

Pricing Extension of Time overheads using the Contract Preliminaries

Some Contracts may breakdown the project into rates, (or prices) for the individual tasks, or sometimes a price for achieving particular milestones. In addition there is often a separate price for the contractor's Preliminaries which are the contractor's overhead costs (or costs to manage the project). The Preliminary price could be further split into Fixed Preliminaries, Value Related Preliminaries and Time Related Preliminaries.

> For example: a project could have a total value of $1,000,000 which is made up of $700,000 for the measured work including mark-up for this portion, $50,000 for Fixed Preliminaries, $50,000 for Value Related Preliminaries and $200,000 for Time Related Preliminaries.
>
> If the project was 8 months in duration the contractor is normally reimbursed the Time Related Preliminaries every month at the average rate which would be $200,000 divided by 8 months which is $25,000 per month
>
> If the contract is extended by 1 month many employers would accept the Extension of Time costs as 1 month multiplied by the average rate, which is $25,000.

The danger with this approach for the contractor is that their actual monthly running overhead costs are not constant throughout the duration of the project. Usually, plotting the actual running overhead costs against time we see a bell shape. The running overhead costs build up slowly from the start to reach a peak when there's maximum activity on the project and then the costs decrease sharply towards the end of the project when the contractor is finishing the work and conducting commissioning. Therefore at the start of the project as the contractor is still in the process of reaching full mobilisation their monthly overhead running costs would be less than the average Preliminary costs evenly spread over the eight months. Their costs would also be less than the average towards the end of the project as they demobilise and reduce staff and equipment. While at peak production, in the middle part of the project, the contractor's overhead running costs exceed the average. From this we can deduce that if the delay is near the start or end of the contract and the contractor is reimbursed their overheads at the average monthly costs they'll probably receive more than the actual costs they incurred. But, if the delay is towards the middle of the project, at peak production, their actual running overhead costs will exceed the average rate which they are being reimbursed.

There are therefore risks to both the contractor and the employer if the contractor is paid Extension of Time overhead costs at the average rate of the Preliminaries in the Contract Document. Furthermore the contractor may be further disadvantaged if when they priced the project they didn't accurately include all their overhead running costs. If some costs went into the Value Related overhead costs these won't be paid unless the actual value of the project increases.

Costs to include in Variation Claims

The contractor should include all of the legitimate and claimable costs in their Variation Claim. It's usually difficult to add in extra forgotten costs after the Claim is submitted. It's unprofessional and will annoy the employer. It may even cast doubts on the legitimacy of the original Claim.

Unfortunately some contractors don't price Variations correctly which means that they could incur costs for which they aren't reimbursed – resulting in them losing money. Other contractors try to be too clever, focussing on how they can make their Claim as large as possible, even inventing costs. Yet, these same contractors forget to recover some of the actual costs they are entitled to claim.

Variations could include:
1. Construction labour costs including:
    a. The base wages.
    b. Overtime costs where applicable.
    c. Non-productive time such as paid breaks, paid travel time, time to prepare hazard assessments, and time to attend inductions.
    d. Special allowances.
    e. Statutory levies such as for training.
    f. Leave pay, bonuses, sick leave, and pension contributions.
    g. Personal protective clothing.
    h. Personal small tools.
    i. Accommodation (if applicable).
    j. Travel (where necessary).
2. Construction material costs including:
    a. The actual material cost (which should be market related).

b. Transport of the material to the construction site.
c. Offloading and handling the material.
d. Protection and packaging of the material (if this is extra).
e. Quality procedures and tests, where required.
f. Wastage due to breakages and cutting. (It should be noted the contractor can't claim for wastage and breakages due to ordering materials incorrectly, or their carelessness with handling and installing the material.) However there is often an accepted industry allowance for breakages and cutting, or the contractor may have to show the build-up of similar rates included in their original price to demonstrate this allowance.
g. Cutting (unless this has already been included in the labour cost).
h. Fixings. This could include bolts, adhesives and welding rods.
i. Royalties.
j. Insurances.
k. Duties and taxes.
l. Preparation of shop drawings and templates where required and when they are an additional cost.
m. Allowing for compaction factors. (For example contractors may purchase road base materials and pay for it per ton or per cubic metre of delivered material. However, the employer usually pays for the item per cubic metre placed in the project – which is after it has been compacted and reduced in volume.)

3. Construction equipment including:
   a. Hire or lease costs. Sometime the contractor uses their own equipment and in this case they would have to demonstrate that the costs for this equipment is market related, or they would have to provide a full breakdown of their costs showing the purchase price, depreciation and maintenance costs.
   b. Unproductive time.
   c. Mobilisation and demobilisation costs if the items have to be brought especially to the project to carry out the Variation work. The contractor may have to demonstrate that suitable equipment wasn't available on the project site or available at a nearer location from where they mobilised it from.
   d. Insurances.
   e. Fuels and lubricants.
   f. Wearing parts such as cutting edges, and drill bits and moil points.

g. Maintenance.
h. Supporting vehicles such as fuel and service vehicles.
i. Attachments and ancillary items.

Some of these costs could be obtained from the equipment manufacture's handbooks or from industry standard publications. Alternatively the contractor needs to keep accurate records of all costs.

4. Demolishing existing structures including:
    a. Loading and transporting of the demolished waste materials.
    b. Temporary supports and bracings.
    c. Access scaffold where required.
    d. Dump fees.
    e. Protection of finished work while the demolition work is executed.
    f. Locating, protecting and disconnecting existing services which are impacted by the demolitions.
5. Supervision and management costs including:
    a. The basic salary.
    b. Allowances.
    c. Leave pay and bonuses.
    d. Accommodation and transport where this is paid by the contractor.
    e. Computers, radios and mobile phones. (Computer costs may also include software costs.)
6. Off-site staff such as Contract Administrators, and Planners
7. Project insurances and sureties which may have to be amended to take into account the increased Contract Value or to take into account varied and additional risks.
8. Costs of permits.
9. Access equipment.
10. Office and facility hire.
11. Additional security if required.
12. Protection of existing and completed structures.
13. Additional design and drawing costs.
14. Subcontractors' costs.
15. Temporary works.
16. Profits and overheads on the above costs.

Although it's important to ensure that all costs are captured it's also petty to be claiming for a kilogram of nails on a one thousand dollar Variation, or to be adding in every cent, which can also be time consuming. Some of the costs may also already be covered elsewhere in the Contract. Remember too that the employer will often request a breakdown of how the contractor calculated the cost of the Variation so it must be possible to justify all of the costs. Good record keeping is an essential aid to proving these costs.

Disruption costs

Disruption costs include:
1. Loss of productivity which results from a number of factors including:
   a. Trade stacking which is caused when a number of trades have to work in an area at the same time. Workers often have to wait for other trades, work in confined or congested areas or around other trades. Their newly completed work could be damaged by these other trades before it's had a chance to 'set' or dry resulting in rework.
   b. Loss of morale which is caused when workers are moved around in what they perceive to be a random fashion, or when completed work has to be redone because the employer issues a revised drawing or Instruction. Loss of morale also occurs when workers are waiting for information or access and aren't gainfully employed or aren't able to use their skills.

      For example: carpenters are used in other roles because they can't access the area where their skills are required. They may have to do work which isn't related to carpentry.
   c. Workers having to move to new task areas meaning they have to reset-up for the task resulting in lost time while the original work area is tidied-up and other equipment and instructions are obtained for the new task. Workers normally go through a learning curve and gain proficiency on repetitive tasks with consequent improvements in productivity. When this is broken the learning curve begins again reducing productivity.

   Quantifying loss of productivity is difficult. If the Construction Schedule is fully resourced it may be possible to compare the actual resources used with the resourced allowed in the Construction Schedule and claim the additional resources as a disruption cost. However, employers will often dispute this and claim the Construction Schedule was under-resourced or that the extra resources were a result of the contractor's poor management. Without a properly resourced Construction Schedule proving additional costs due to a loss in productivity becomes even more difficult. The contractor may have to prove by other means (such as showing the actual breakdown of their price) what the activity should have cost and compare that to the actual costs. Some employers may accept a percentage loss, but unless there's some concrete basis for this percentage contractors will face difficulties. In some cases the contractor may have to rely on expert opinion to back their assertions of what normal productivity should be. (See the 'measured mile' method below.)

2. Doing work out of sequence resulting in equipment or personnel having to be remobilised.
3. Having to use different equipment to what was planned. For instance larger cranes might have to be used to install equipment because the work occurs later in the Schedule after other structures have been completed which restrict access. In some cases there might be more 'hand work' because of restricted access caused by completed work which prevents mechanical equipment from accessing the area.
4. Having excessive peaks and troughs of resource usage. Contractors generally try and smooth their manning and equipment requirements to ensure a continuous usage. Peak requirements may necessitate the mobilisation of additional resources for short periods (with the additional costs and disruption of mobilisation), while in the troughs of low requirements some resources may be idle or not fully utilised.

Formulating disruption costs using the 'measured mile'

As mentioned in the previous section it's difficult to prove the actual costs caused by disruption and delays. One method is known as the 'measured mile'. This compares the actual production (man-hours or equipment hours spent) on the disrupted portion of the project with what could be expected normally. The hours or costs in excess of those considered normal would then be considered as the disruption costs. There are a number of methods that can be taken to derive the costs deemed to be normal, or the 'measured mile'. These could be:

1. If the contractor has completed a similar set of tasks on the same project and has a record of the hours expended in completing these task, they could be compared to the hours expended on the disrupted portion to formulate the disruption costs.
2. If the contractor has completed a similar project the hours expended on that project could be compared to the hours expended on the project that's been disrupted.

3. The contractor employs an independent expert industry witness who can produce records of the hours or costs of a similar type and size project which can then be compared with those expended on the disrupted work.
4. If a portion of the project was disrupted then by considering the hours expended on the project before disruption and comparing them to the hours expended during the disruption.

> *For example: if the contractor used 5,000 man-hours to complete 40% of the project before disruption and then spent 10,000 man-hours to complete the remaining 60%. Using the initial 5,000 man-hours and extrapolating this then the remaining 60% should only have taken 7,500 man-hours so the disruption cost an additional 2,500 man-hours.*

It should be noted that the measured mile is an evaluation of damages and doesn't have to be exact, but it's considered a reliable and acceptable approach. However, where the contractor has suffered additional inefficiencies due to an event caused by them, then the hours expended on this incident should be deducted from the calculations.

> *For example. in the above example If the contractor had a quality problem that meant they had to demolish and rebuild a structure they would need to calculate the hours to undertake this work and deduct these from the 2,500 hours calculated above to arrive at the actual hours directly attributable to the disruption.*

Overhead and profit on Variation costs

Contractors would obviously like to add on as much profit and overhead as possible to the costs they are entitled to claim in their Variation. However, this amount, which is normally expressed as a percentage, must be justifiable, fair and reasonable and where possible within the norm in the industry. Often in the request to price or quote for the project the contractor is asked to submit the overhead and profit percentage that will be added to Variations and this value is included in the Contract Document. This is the easiest route and contractors should consider adding an item to their quotation, or price, indicating what percentage their profit and overheads will be for Variation work even when the employer hasn't asked for it. If an overhead and profit percentage isn't included in the Contract Document the contractor may have to use an industry norm which might be 10% and they would have to justify this by providing evidence or expert opinion. Alternatively they would have to open their original pricing calculations for the project so the employer can see their calculations and the profit and overhead that the contractor originally added to their costs when they formulated their price.

Inflating Claims

Some contractors automatically inflate the value, or the time requested, in their Claim. There's the 'let's try our luck' attitude and we'll see what we can get and we may get more than we deserve. Then there's the approach; 'whatever we claim the employer is going to bargain us down and want a discount'.

I believe in ensuring that I've included all of my provable costs and delays and presenting my Claim in such a fashion that the employer can't refute it. Once an employer realises that the contractor has 'padded' their Claims (added extraneous and inflated values) they'll automatically review every Claim the contractor submits with suspicion – looking for the extra time and money they know the contractor isn't entitled to ask for but which they always claim.

It should also be remembered that when there's a dispute the Claim will be reviewed by experts that will see through the extra 'padding' resulting in them favouring the contractor in a poor light. Sometimes adjudication of Claims can go either way. It won't help the contractor's Claim if they have annoyed the adjudicator by presenting a false Claim or adding in time or values that can't be substantiated. Nobody wants to find that they've been deliberately misled and had to waste time uncovering false information.

Cost calculations

It's important for the contractor to keep all their calculations together so they can easily be referred to. Sometimes we tend to put numbers down without referencing where they came from, then, when the employer later asks we can't remember how we derived the number, occasionally even providing a slightly different number. Often in the course of a project there are a number of different Variation Claims with similar costs and calculations. So for instance, if we are using a shift rate for a general worker we must be able to clearly demonstrate how we developed the rate we are claiming, which is made up of their base rate plus leave pay plus overtime plus sick benefits plus other benefits, etc. This same rate should be consistently used in all other Variation Claims which include general workers. Differing costs, or cost calculations, will be viewed suspiciously by employers, and suspicion can ultimately lead to rejection of the contractor's Claims

Employers will ask questions. They will want to see how costs and prices have been arrived at. Contractors must be able to convince their employers that the costs used to develop the Variation costs are correct and not some 'thumb suck' to get money from the employer.

Have the full impacts of the Variation been considered?

Often Variations impact the Schedule, even though this isn't always obvious. It's therefore important for contractors to assess the effect the Variation has on the Schedule which may result in:
1. The overall project duration being extended.
2. The Variation impeding or delaying other works.
3. The Variation work requiring resources from other tasks which then impacts on the completion dates of those tasks.
4. In some cases the overall project may be delayed by the Variation extending the project into an unfavourable weather season such as winter or periods of heavy rain.

Sometimes one Variation can have a knock-on impact on other tasks so the contractor must ensure there are no unintended consequences. The contractor should also check with subcontractors that they aren't impacted.

For example: a request to build a wall a metre away from its original intended position may seem simple. But there could be knock-on effects on plumbing pipes or electrical conduits that have already been installed on the basis the wall will be in its original position.

Pricing the Variation using existing prices

Sometimes employers insist Variations are priced according to existing prices in the Contract Document for other similar work. Contractors then have to ensure that the existing price will cover all of their costs. If they don't then the contractor will need to justify to the employer reasons why the price in the Contract shouldn't be used.

When using existing prices in the Contract contractors should check:
1. The work is the same as the work priced in their original quotation.
2. The access to the work area is the same as was allowed for in the original price.

For example: I've had employers introduce minor variations part way through the project which could be as simple as adding in a small brick wall (a couple of square metres) in an apartment on the fifth floor long after we've completed the brickwork on that floor. Employers think this should be done at the same rate as the brickwork for the rest of the building. They don't realise that the access is now no longer the same. The materials probably have to be moved by hand through areas which have already been completed. The team doing the brickwork has to be especially mobilised to that area. The production rate to build that small wall could be as low as 25% compared to that to build the rest of the building. Obviously in this case the same rate cannot apply.

3. That the materials are still available for the same price. Maybe a small quantity has to be especially ordered to do the work which could incur additional purchase and transport costs. In the meantime the cost of the materials may have increased – a problem that often occurs with projects with a longer duration.

4. That the equipment, subcontractors or skills required to do the work are still available on the project. If these have been demobilised they will have to be remobilised at additional costs, and often at a more expensive price.

Pricing the Variation using Cost-plus methods

In some cases the employer may issue an Instruction for additional work to be carried out on a cost-plus, or cost reimbursable, or proven cost, basis. Basically the contractor must produce evidence of the actual costs incurred and charge these to the employer. This may seem a simple exercise and a Variation where the contractor cannot lose money. Unfortunately it's not always as straight forward as it seems and if the contractor is not careful they may find that they incur costs which aren't reimbursed.

Before undertaking work on this basis it's important to agree:
1. The basis of the charges, such as what the employer will pay for and what is assumed to be included in the cost. For instance what's included in the cost for an item of equipment (such as fuel, maintenance, insurance, etc.) and how is overtime treated.
2. How the contractor will be reimbursed for unproductive time (for example moving personnel and equipment between tasks).
3. How the contractor will be recompensed for their site overheads, facilities, supervision and management.
4. Who, from the employer has the authority to agree and sign for the hours worked.
5. What the employer will be providing.
6. What the contractor's mark-up is on these costs.

7. The type of records and proof required by the employer to substantiate the costs.
8. If there's an overall maximum value that shouldn't be exceeded.

It is important that accurate records are kept of all costs and hours worked. These costs may include amongst others:
1. Transport of personnel, materials and equipment.
2. Handling and off-loading of materials.
3. Taxes.
4. Wastage and breakages.
5. Insurances.
6. Royalties.
7. Mobilisations and inductions.
8. Safety equipment and clothing.

Hourly records of the people and equipment used to do the work should be signed daily by the employer's representative.

It's easier to track the costs of the materials if these are ordered in specifically for the task and the invoices and delivery notes can be kept as supporting documents to validate the costs charged to the employer. However, often materials are drawn from the contractor's store or taken from other tasks. If accurate records aren't kept of these materials they might not be charged to the employer. Also, invoices will have to be tracked down for the original deliveries to prove the cost of the material. In some cases the material ordered to replace the stocks that they were drawn from might be more expensive than the original purchase invoices which could cause the contractor to be out of pocket since they claimed from the invoices for material delivered earlier for other tasks.

It should be noted that the charges for materials should be fair and reasonable and the employer may reject invoices for materials should they believe the contractor hasn't procured the items at a market related rate.

The contractor's staff undertaking the work must understand how the employer will be paying for the work and what items they should be recording.

Records of people and equipment should agree with the numbers recorded in the daily diaries and weekly reports to prevent confusion, and the possibility the contractor is reimbursed for a lower number of people.

If there is a maximum value for the Contract or Variation Order it should not be exceeded and the employer should be advised well before this figure is reached.

It should be noted that this method of pricing shouldn't be looked on as an 'open cheque' where the contractor can mismanage the work and still be paid. Employers may dispute costs where the contractor has been seen to waste materials or where it can be proven that the contractor's productivity wasn't acceptable.

Validity of the price

In some instances, employers request contractors to price a Variation before the contractor can start the work associated with the Variation. Once submitted the employer may take several days or weeks before they accept and approve the Variation. In the meantime work is probably continuing on the project. By the time the employer accepts the Variation and issues a Variation Order the contractor may have executed work that now has to be demolished to make way for the Variation work they priced – demolitions which weren't allowed in their original price. Alternatively, work could have advanced which now blocks or limits access to execute the Variation, or the equipment that was available on the project is no longer available and equipment will have to be mobilised – but costs for mobilisation weren't allowed in the price.

It's therefore important to ensure that the Variation costs stipulate any assumptions that were made in the contractor's pricing, what has been allowed in their price and any factors which might change the price or make the price invalid. If necessary contractors should advise their employer that time is of the essence in approving the Variation – even imposing a time limit for the validity of their price.

Supporting Documentation

It's important the Claim includes all supporting documentation. Depending on the Claim these documents could include:
1. The approved Construction Schedule.
2. The impacted Construction Schedule – showing the effects of the delay events.
3. The Information Schedule.
4. Relevant clauses from the Contract Document.
5. Extracts from the Daily Diary or Log.
6. Drawing Issue Logs.
7. The Drawing Register.
8. Correspondence (letters and emails).
9. Actual weather records for the project as well as those from the nearest weather station showing the average recorded conditions for the location.
10. Photographs.

Chapter 3 – Preparing the Claim | 97

11. Minutes of meetings.
12. Copies of Instructions.
13. Copies of Requests for Information.
14. Material invoices.
15. Contract Notices.
16. Calculations.
17. Relevant clauses from the specifications.

These documents need to be included in a logical order and should be referred to in the Variation Claim. It may be necessary to include the supporting documents as appendixes. The Variation Claim shouldn't include irrelevant documents.

Contradictory documents should be explained. The employer shouldn't uncover unexplained contradictory documents which could weaken the contractor's arguments.

Checking the Claim submission

Before submitting the Claim the contractor should discuss it with the construction team, particularly those that are involved with the work relating to the Claim. It's possible they may be able to add to the Claim, notice items and information that have been overlooked and supply extra costs or supporting information that could be included.

Someone else should check the Claim for errors and to see if they can follow the logic. If the logic is flawed, or can't be easily followed, employers might not understand the Claim, possibly be confused and inclined to reject the Claim first, before asking questions. Once a Variation Claim is rejected it becomes more difficult to change the employers mind and convince them that the contractor is entitled to the Variation.

Contractors should carefully check the submission for arithmetic errors. An error could mean that the Claim amounts to a lessor value to that which the contractor is entitled to. If the employer finds an error that's detrimental to them they might become doubtful of the facts and figures making up the Claim.

Ensure the Claim includes for all delays and costs. Once submitted it is awkward to add in extra delays and costs to the same Claim. Adding additional costs and delays later often creates confusion and casts doubts over the legitimacy of the original Claim. After the Claim has been approved by the employer it's almost impossible to go back to the employer and ask for more.

Get the Claim right first time.

Why is the Claim so expensive?

Sometimes what appears to be a small Claim results in a comparatively large value, or an inordinately large delay. In these cases it may be pertinent to explain to the employer why the value, or the delay, is larger than expected. Employers often don't understand why such a large quantum is being claimed by the contractor and may think the contractor is requesting more than they are entitled to. Some employers have been known to reject Claims just because they felt the Claim was overly excessive, without giving due regard to whether the contractor was entitled to submit the Claim, let alone whether the amount requested was correct.

Termination

Reasons for Termination should be listed in the Contract Document. The contractor must ensure the reason for Termination is valid. They should decide whether Termination is because it is impossible for the project to continue, or whether it's due to the other party not fulfilling their terms of the Contract. It should be noted, that in some cases the Breach could be straight forward such as if one of the parties becomes insolvent. However, the contractor's poor performance leading to slippage on the Schedule is probably not a reason to Terminate, especially if the Contract Completion Date is some way off and the contractor still has time to finish the project by the specified date.

Notice provisions in the Contract must be strictly followed. Normally there's a requirement to submit a written notice to the specified address in the Contract Document. A signed receipt should be obtained. This notice should include:
1. A specific statement that the party is in Breach
2. Specify the nature of the Breach citing the correct contractual Termination clause.
3. Outline the events leading up to this notice specifying correspondence and meeting minutes raising concerns relating to the event.
4. Provide a time period by when the Breach must be rectified.

If the Breach is not rectified within the specified time period a second notice Terminating the Contract must be delivered in writing to the person specified in the Contract. The notice must clearly specify that the Contract is being Terminated, and the reasons for the Termination, citing the Termination Notice previously sent.

Failure to follow these procedures correctly may give the other party reason to Terminate the Contract for wrongful Termination by the other party. They can then claim damages.

It should be noted that Termination makes all actions in terms of the Contract null and void after Termination. In other words, if the contract is Terminated before the Completion Date the contractor cannot be held liable for Liquidated Damages because the Contract Completion Date will occur after the Contract was Terminated. All actions in terms of the contract which occurred before Termination still have to be complied with. Work completed by the contractor before Termination must be paid for in terms of the Contract.

As mentioned in Chapter 1 both parties need to accept a Repudiatory Termination to validate it. If a Repudiatory Breach is accepted the party suing for Termination can claim damages to an amount that would compensate them as if the Contract was completed. In other words they could claim the additional costs to appoint another contractor to complete the project.

Insurance Claims

When an insurable event occurs it's important that the relevant insurance company is notified as soon as possible. The contractor should know if the event is insurable, under what policy it can be claimed, and who the insurer is, and what the excess is. Photographs should be taken of the damage, and the area made safe and secured. A detailed report must be prepared and submitted to the insurer. This report should include:
1. The insurance policy number.
2. The date and time of the event.
3. The location of the project.
4. What the event was.
5. What the damage is
6. An estimate of the repair costs.
7. Mitigating actions taken, if applicable.
8. Photographs of the event and subsequent damage.
9. Any injuries.
10. The repairs implemented, or actions taken, to secure the facility and make it safe.
11. Names of witnesses if applicable.
12. Contact details of the person submitting the report.
13. The urgency of the response required from the Insurer.

The employer should be immediately advised if the facility and construction works has been damaged. The owners of equipment damaged in the event need to be advised straightaway. The contractor should keep a record of who was advised of the damage and the applicable case numbers.

In certain incidents the police will have to be notified – again the contractor must obtain a case number. Certain accidents on the project are notifiable to the authorities.

When the insurer is contacted they will normally advise if an assessor will inspect the damage and what the next steps are to repair it. Only once the insurance assessor gives the go ahead to proceed can the damage be repaired.

Obviously time is of the essence since the damage may be delaying progress on the project, and normally most insurance policies will not cover the impact of these delays. It's important the insurer understands how critical it is for repairs to be completed.

Insurance Claims should include all of the costs associated with carrying out the repairs and making good damage including:

1. The cost of the materials together with transport, insurance and handling.
2. All labour costs.
3. Costs of cleaning, removing and disposing of debris, including pumping out storm and flood water.
4. Protection of the undamaged work which is affected by the repair work.
5. Demolition of damaged structures.
6. Supervision and overhead costs.
7. Damage to plant and equipment. It should be noted that in many instances these damages may have to be claimed under the individual policies of the items.
8. Temporary support or access structures.
9. Subcontractors' costs associated with the damage.

Unfortunately most insurance policies don't cover for consequential damages. Any delays caused by the event, including their resultant costs, won't be covered.

Claims from subcontractors

There are normally two types of Claims contractors receive from their subcontractors:

1. Claims directly attributable to the contractor's employer.
2. Claims that are solely due to the contractor delaying the subcontractor, or making changes to the subcontractor's Scope not directly attributable to the employer.

Sometimes subcontractors combine Claims that are due to the employer and to the contractor. Some Claims may be due to both parties and it's necessary for the contractor to apportion the Claim correctly.

Chapter 3 – Preparing the Claim | 101

Some contractors simply pass on their subcontractor's Variation Claim directly to their employer without checking that the:

1. Claim is attributable to the employer.
2. Arithmetic is correct.
3. Subcontractor has used the approved prices in the Contract Document.
4. Claim is clearly presented.
5. Claim is presented on the contractor's own letterheads.
6. Contractor's profit and overheads have been added to the Claim.
7. Claim includes all of the subcontractor's entitled costs.
8. Claim includes any costs the contractor incurred in supporting the subcontractor to execute the work associated with the Variation.

Many subcontractors are unsophisticated when it comes to developing and presenting Claims. It's often advisable that contractors assist their smaller subcontractors in preparing their Claims attributable to the employer. This has a number of benefits which includes:

1. Poorly presented Claims can irritate and annoy the employer.
2. Ensuring the Claim is maximised, since usually the contractor can add their profit and overhead on top of their subcontractor's Claim, so the larger the subcontractor's Claim the larger the profit and overhead that can be added.
3. The more money the subcontractor can Claim from the employer the more likely they are to be profitable. A profitable subcontractor is usually a happy subcontractor that is often more willing to help the contractor on the project and to price future projects for the contractor.

Claims that are solely a result of the contractor's actions (or lack thereof) need to be assessed to see that they are legitimate and that the prices used are correct. It's important to deal with the subcontractor's Claims in a timely fashion so they don't become a cause of friction which could negatively impact the subcontractor's performance and possibly their cash flow. Dealing with the Claim immediately it arises also means that the people who have knowledge of the Claim and the events surrounding the Claim are still available to provide the background and history to the Claim. Once the Claim is dealt with both the contractor and subcontractor have certainty with their project budgets.

Summary

1. Every Claim should be carefully prepared by an experienced member of the contractor's team who is familiar with the project.

2. The Claim should take into account all of the costs and delay impacts.
3. Claims need to be lodged in accordance with the Contract and within the time-frames of the Contract.
4. The Claim should detail the cause of the event, correspondence and events leading up to the event, the implications of the event and the impact of the event. The Claim must be well thought through and presented so that it is understandable to others.
5. Claims should be checked to ensure there are no arithmetic errors and that all costs and delay impacts have been included.
6. Claims must include, or refer to, all the supporting documentation.

Chapter 4 – Negotiating and Winning Claims

As much as we all consider our Claim as complete, final and non-negotiable when we submit it, it's almost certain that the employer will come back with questions, counter-arguments and try and negotiate the quantum of the Claim down, or even reject it completely.

In responding to questions from the employer the contractor must ensure that replies remain factual and stick to the original arguments presented. It's easy to introduce other arguments and ideas which could be contradictory or confusing.

Questions must be replied to in writing referencing the original Claim submission. If the questions are raised in a meeting the contractor should take notes of the questions and their answers, and confirm them afterwards in writing.

It goes without saying that all correspondence relating to the Claim needs to be filed with the other documents related to the Claim, and where it's readily able to be found and retrieved. Some Claims can take months to resolve, and if the Claim follows the legal or arbitration route it could even take years.

Why disputes can be harmful

Where possible disputes should be avoided because they:
1. Are time consuming – contractors shouldn't underestimate the time that disputes take to resolve, time that could be better utilised elsewhere on the construction project, or on other projects.
2. Can damage the construction company's reputation – employers and subcontractors often avoid contractors that have a reputation for project disputes
3. Damage the relationship between the parties – the relationship seldom returns to normal after a dispute.
4. They are costly especially when they become legal and involve lawyers.
5. They may end poorly for the contractor who doesn't receive the full value of their claim.

6. They can disrupt the construction project and sometimes cause further problems.

How to avoid disputes

Submitting a Variation Claim is not a dispute. A dispute arises when the contractor and the employer cannot agree that the Variation Claim is legitimate or they cannot agree the quantum of the Claim. Normally before a dispute is declared the claimant would have submitted a Variation Claim and the respondent would have replied either rejecting the Claim in full, agreeing to part of the Claim or asking for further information. The claimant would supply the requested information and could respond with further back-up documentation if it appears the respondent has misunderstood the Claim or rejected it unfairly. Often a meeting can answer many questions allowing both parties to understand the other's reasoning. Only when it appears that the parties cannot agree on the Variation Claim does it become a dispute.

Generally most disputes can be avoided if appropriate actions are taken, such as ensuring:
1. There's a legally enforceable Contract in place which protects both parties' interests.
2. The Contract is well written and doesn't have conflicting clauses or contractual loopholes.
3. The contractor understands the Contract and complies with its provisions. Often the contractor's Project Manager doesn't read the Contract until a dispute arises – it's normally too late then!
4. The contractor communicates with the employer and their subcontractors, timeously notifying them when problems, delays and Variations arise.
5. The contractor submits and resolves Variations as soon as practical.
6. Employers act in a fair and reasonable manner.
7. Accurate records are maintained.
8. There's willingness by the parties to talk and negotiate.
9. Personalities and emotions are kept out of the negotiations.
10. The contractor admits when they're wrong and doesn't lodge extravagant Claims.
11. The consequences of escalating the dispute are weighed up carefully, since the costs of legal action may be more than the outcome is worth.
12. The construction Contract is administered in a spirit of honesty and cooperation by all parties.

13. Senior management are made aware of potential disputes and problems on a project, thus enabling them to take the necessary action and intervene if required to avoid the problem escalating.
14. Expert advice is sought when necessary.
15. The contractor deals with their subcontractors and suppliers fairly and resolves Claims when they arise.

The cup of coffee approach

It's often helpful for the contractor and the employer to meet informally, possibly over a cup of coffee. These meetings are an opportunity for each side to explain their frustrations and where they believe the other party could do better.

It's also an opportunity for the contractor to provide details of Variation Claims that are about to be submitted, the rationale behind them and the background to how they arose. This explanation ensures the employer isn't surprised when the Claim arrives on their desk. By understanding the reasoning behind the Claim the employer may be more receptive to the contents of the Claim.

Informal discussions around a Claim that can't be resolved can produce suggestions for 'unblocking' it, and create a way out of the deadlock. Listening to the other side's concerns, restraints and point of view often gives a better understanding of their arguments. However, contractors must remember that these are informal discussions and should be careful not to make agreements, or accept commitments from the employer, which aren't recorded. When an agreement is reached at one of these meetings it should be confirmed in writing.

The important thing is that by talking to each other the parties can cut short what can sometimes be an acrimonious trail of correspondence and emails. Talking to each other can often produce a way forward where both the employer and contractor can be winners. Talking can allay the fears the parties may have about each other's conduct and remove any distrust.

Employers that don't respond to Variation Claims

Unfortunately some employers take the approach of not responding to Variation Claims. This may be because the person administering the Contract on behalf of the employer doesn't want to raise the Claim with their manager or their employer because it could reflect poorly on their project management skills. Some employers hope that by leaving the Claim it may just go away, or that they can take all of the Claims at the end of the project, put them in one 'basket' and negotiate a once-off settlement with the contractor – a contractor who at that stage will be desperate just to get some money from their Claims. Alternatively employers may in the meantime put together their own counter Claims against the contractor in the hope that they can trade these Claims off against the contractor's Claims.

This approach is dishonest. Unfortunately most Contracts specify a time for the contractor to issue a Variation Claim, but very few specify a time by when the employer has to respond to a Claim.

This approach is unfair to the contractor since if the Claim isn't approved it won't be paid which disrupts their cash flow. It also leaves them in limbo not knowing if they'll be paid for the work, which means they could be facing a project loss which can, in the case of large Claims, have a devastating impact on the company.

Furthermore unresolved Extension of Time Claims may mean that the contractor is faced with Liquidated Damages because they cannot complete the project by the original Completion Date. These Damages can further harm the contractor's cash flow. The contractor faces further uncertainty not knowing if they should be Accelerating at their cost.

Also, as long as Claims aren't resolved there's a risk that the employer's budget may be overspent when the Claims are finally resolved, which could result in the contractor not being paid for these agreed Claims.

Contractors need to make every effort to get the employer to resolve Claims as soon as possible. Obviously, ensuring they have provided all of the supporting information when the Claim is first presented is essential, and also that all of the employer's questions are answered as speedily as possible. A list of Variation Claims, with their anticipated value or time impact, should be included at all project meetings with the employer. This should show those Claims which have been agreed and those which are still unapproved. This Claims register is also an aid for the employer so they can adequately monitor their project budget.

Should the employer be unresponsive to Variation Claims, then the contractor should submit letters asking for a resolution. Further action could involve requesting a meeting with the employer's representative and even, if necessary, taking the problem to the employer's senior management who might not be informed of the status of Claims. In some instance the contractor may be dealing with the employer's Agent (the Engineer, Architect or Project Manager) and the employer may not even be aware of the unresolved Variation Claims. In these instances it may help to bring the problem to the attention of the employer and ask for a resolution to the problem. If Claims remain unresolved because they aren't replied to it may be necessary for the contractor to follow the dispute resolution process outlined in the Contract Document.

Contractors shouldn't let Variation Claims go unresolved until the end of the project as they risk not being paid the Claim. Furthermore, often later Variation Claims are dependent on attaining a resolution for an earlier Claim. As long as the first Claims aren't resolved contractors have to base their later Claims on the assumption that their first Claims were unsuccessful or didn't exist. This becomes messy and creates confusion with the final value and quantum of the delay uncertain. When the first Claims are resolved it may require the later Claims to be reworked taking the earlier Claims into account, then resubmitted, which takes time and causes confusion.

The impartiality of the employer's Agent

Employers often appoint an Agent who is responsible for administering the Contract, issuing Instructions, ensuring the project is completed on schedule and in accordance with the employer's brief and Scope of Work, and that the project meets the required quality expectations and specifications. Their role is also to protect the employer's interests and to assess and adjudicate Variation Claims. This Agent is often the Engineer, Architect or the employer's Project Manager.

The Agent is expected to assess Claims in a fair and impartial way, protecting the employer's interests while being fair to the other parties. Unfortunately, often Variation Claims arise because of a fault caused by the Agent acting in their role as Engineer, Architect or Project Manager. Sometimes in these cases the Agent fails to act impartially and rejects the contractor's legitimate Claim because the Claim exposes the Agent's shortcomings which caused the Variation Claim. Indeed some Agents go to extreme lengths to either refute the contractor's Claim when they themselves have caused it, or to point the contractor in another direction, indicating they'll approve a Variation Claim that doesn't implicate them. Rejecting the contractor's Claim on these grounds is dishonest and often leads to costly disputes.

Should the Agent request the contractor to phrase the Variation in different words, taking a different view on the Claim so that the Agent and their company isn't implicated, but with the understanding the Variation Claim will then be approved, puts the contractor in a difficult position. The easy option is to follow the suggested route and resubmit the Variation Claim taking out any reference to the Agent's shortcomings on the understanding that the Agent will approve the Claim even though the revised Claim may now be tenuous and have lost its legitimacy. The risk is that on review the employer, or an employer appointed auditor, later rejects the Claim because it is exposed for what it now is, a bogus Claim. If the Claim that's resubmitted in accordance with the Agent's wishes is rejected it will be difficult to resubmit the original legitimate Claim, or to point a finger at the Agent saying the Claim was resubmitted in accordance with their wishes.

Contractors have to decide whether they'll face down the Agent and declare a dispute, or whether to oblige the Agent and resubmit their Claim. The choice may be influenced by the size of the Claim and on how legitimate the resubmission will be. Often when there is more than one reason for submitting the Variation Claim, both having a similar result, it may just be simpler to select the less contentious reason.

Contractors should avoid pricing projects when they know the Agent won't act in a fair and impartial manner.

Claim Registers

Contractors should maintain a Claim Register, or log, which should include:
1. The date when the Claim was first discovered.
2. The cause of the Claim.
3. The date the employer was notified of the contractor's intention to lodge the Claim.
4. The date the quantum (time and/or value) of the Claim was submitted to the employer.
5. Dates when the employer responded to the Claim.
6. If the work in the Claim has been executed.
7. If the Claim has been resolved and the final agreed quantum of the Claim.
8. If the employer has issued a Variation Order to cover the extra value.

This register is a reminder to the contractor of Claims that have to be submitted as well as Claims that are unresolved.

The Claims Register should also be included with the project site meeting (between the employer and the contractor) minutes. It not only serves as a reminder of the outstanding Claims, but is also an aid to the employer to ensure they have sufficient budget for the remaining works.

Variation Claim meetings

Once the Variation Claim has been submitted the employer may respond in writing to the contractor with comments, queries and sometimes agreeing or refuting the Claim. Often though, at some stage the employer will call a meeting to discuss the Claim or Claims. In fact I recommend that contractors try and persuade their employers to call a meeting sooner than later. Sending correspondence backwards and forwards can mean it takes longer to resolve the Claim. In addition the letters take time to draft. Miscommunication can result in confusion. Once the employer has rejected a Claim in writing it's sometimes difficult to persuade them to reopen the Claim and change their mind. But, a verbal rejection in a meeting still provides the contractor time to explain why the employer is wrong to reject the Claim. Meetings also provide an opportunity to read the employer's mood and attitude. There may be small clues to what the employer will and won't accept. Meetings also provide an opportunity to negotiate an outcome that will be best for all parties.

Unfortunately sometimes contractors lose the initiative at these meetings because they aren't well prepared or make a rushed decision. It's therefore wise that:

1. At least two people from the contractor attend the meeting. One person is more likely to be bullied and intimidated by the other side, particularly if the other side is represented by a group of people. It also provides a witness to what was agreed. Two or more people can discuss the employer's attitude afterwards, gaining a better understanding and interpretation of the employer's mood and their expectations.

2. All the contractor's representatives understand the basis of the Claim before the meeting. Unfortunately senior managers sometimes attend these meetings and they aren't always fully acquainted with the Claim resulting in them saying the wrong thing, making contradictory statements or agreeing to an unfavourable settlement. Sometimes the Claims have been submitted months before so it's good practise to reread the Claims file to refresh everyone's memory.

3. The contractor has all the documents relating to the Claim with them, and that these are filed in order and easily accessible. If the employer sees the contractor is well prepared with all the documents to present a logical case they'll be more believable than if they can't find documents or are scratching around for missing documents.

4. The contractor understands who is representing the employer and what powers they have. Sometimes it's pointless striking a deal at the meeting if the employer's representatives aren't empowered to accept the deal. Employers sometimes send their minions to these meetings in an attempt to whittle the contractor's Claim down. At a second meeting the employer may then try and negotiate the previously 'agreed' Claim value even lower.

5. Notes are taken during the meeting, particularly of questions that need to be answered afterwards and points that were agreed to.

6. These notes are confirmed in writing (in the form of a letter or meeting minutes) to the employer afterwards so there is a written record of what was agreed and what questions require answering.
7. The contractor's team has an agreed 'bottom-line' number or value they are willing to accept to settle the Claim. Sometimes to avoid lengthy disputes it's worth settling a Claim for a lesser amount than originally claimed, but still an amount where the contractor will be profitable. Employers also want to feel like winners at these meetings so it may be necessary for some 'give-and-take' from both sides.
8. Emotions are kept out of the arguments. Everyone should stay calm and present the case logically, keeping to the facts.
9. The contractor isn't swayed by emotional arguments made by the employer. They may try and beat the contractor up by claiming they've been fair while the contractor is now being unfair.
10. The contractor isn't rushed to make a decision. If in doubt ask for a ten minute break to discuss the issue with colleagues. However, if a meeting is going favourably it's often best to strike a deal on the day, rather than delay the decision to another day when moods may have changed.
11. The meeting is summed up at the end, including what was agreed and what questions the parties still need to answer.

Negotiating a Claim

Being able to negotiate is a skill that's essential to successfully agreeing Claims. There are many useful courses about negotiations and the contractor's team involved with negotiating Contracts and Variation Claims should attend one of these. Negotiating is often a slow process and the success depends on:
1. Understanding the strengths and the weaknesses of the Claim.
2. Being able to support the strengths of the Claim with relevant documentation.
3. Being able to defend the weak points of the Claim, but always ensuring that supporting a weakness isn't done at the expense of credibility.
4. Understanding the other party's arguments and having the facts to explain why they are incorrect. Take time to go through every argument even if you consider the points irrelevant or trivial.
5. Keeping emotions out of arguments. Even if the other party becomes personal or abusive it's important to stay calm and stick to the facts.
6. Presenting arguments in a logical fashion.
7. Being prepared to grant the other party small wins.

Chapter 4 – Negotiating and Winning Claims | 111

8. Understanding the other party's fears or concerns which will enable these fears to be addressed, pointing out why these fears are unnecessary, or presenting arguments that will allay the fears.
9. Learning to admit when you are wrong. Defending a Claim, or part of a Claim, with flimsy arguments wastes time, impacts integrity and frustrates the other party, even undermining any partial wins already negotiated.
10. Knowing what the 'walk-away' point is – what is the absolute minimum that can be accepted. When this point is reached, and in the face of stiff opposition, depending on other factors and Claims it may be advisable to accept the settlement.
11. Understanding the other party's emotions. Some people will launch into blustery and abusive attacks while others may try and appeal to the contractor's sympathy. Often these are made to disarm the contractor. Understand when they are bluffing and when there is serious intent.
12. Knowing the other party's bottom line or budget is helpful as often it is pointless arguing for a figure higher than this because it just might not be achievable.
13. Not making threats or promises that can't be kept.
14. Learning when to take a break with the negotiations when your side needs to regroup or develop a new strategy, or when it's obvious the employer is engrossed with other events, or is just in a 'bad mood'. On the other hand when things are going well it may be worth extending the negotiations so that a deal can be struck while everyone is in agreement.
15. When there is a deadlock, knowing it's pointless arguing the same point over and over again. Repeating the argument isn't going to change minds – nor is raising one's voice. Think of another tack or angle. It may be necessary to leave an issue unresolved and move onto another part of the Claim or discuss other Claims. Settlement of other issues may allow both parties to later approach the deadlocked issue with more open-minds.
16. When a deadlock persists, sum-up the issue, including what points the parties agree on, what the disagreements are, and why the parties differ. Often this summary awakens both parties to how close they are to agreement.

Bribery and corruption

At no stage should contractors be tempted to bribe employers or their representatives to settle Claims. This is illegal, will damage the contractor's reputation and could jeopardise the Claim. Frequently contractors 'wine and dine' and entertain the employer or their Agents, sometimes providing tickets to sporting events and shows. In some cases these tickets may come with additional benefits such as air tickets and hotel accommodation. Some employers have strict rules about handing out and accepting such largess and contractors must take care that they aren't breaking these rules. While the occasional dinner or drink with the employer or their representatives is good for fostering open communication these gatherings should not become too frequent or be construed as a gift or bribe.

Should the employer's manager consider that the contractor has been overly generous towards their Agent, or representative, it could cause questions to be raised, providing reason to reopen and investigate Claims which have already been settled and agreed on suspicion that these were agreed in payment for the contractor's generosity. When large Claims are being negotiated the contractor needs to be particularly careful not to provide gifts, or invite employers and their representatives to events, which could be interpreted as an attempt to favourable influence the settlement of a Claim.

Variation Orders or Contract Amendment Orders

Once the employer has agreed the Variation the norm is for a Variation Order, Contract Amendment Order or Change Order to be issued to the contractor. This document amends the final Contract Price. The Variation Order is confirmation that the value, and (or) the time, for the Variation has been approved. Normally without this Variation Order the contractor won't be paid for the Variation.

When the contractor signs this order they must check that the value and the stipulated time are what have previously been agreed and that no unfavourable clauses have been added. The authorised person from the employer must sign this document to make it legal. The Variation Order should preferably have the value of the Variation as well as the revised approved Contract Value which includes the original Contract Value plus all the previous Variation Orders. These amounts should be carefully checked to ensure there are no errors and that no previously agreed Variations have been omitted.

When disputes are unavoidable

Unfortunately sometimes disputes will arise where Claims cannot be amicably resolved. The employer may have rejected the Claim in its entirety, or they may have agreed to a lesser amount than the contractor believes they are entitled to. The contractor should notify the employer in writing if they aren't in agreement with the employer's ruling on a Claim.

There are various options to resolve disputes.
1. The cheapest and easiest is to negotiate with the individuals directly involved with the project, which may require some compromise from both parties.
2. If negotiation fails the dispute should be referred to senior managers from the different parties. Often the problem is due to a clash of personality or an individual's incorrect interpretation of the Contract and most managers can quickly resolve the issue when these obstacles are removed.
3. If the owner of the facility is not party to the dispute they can be approached to assist in resolving the dispute.

The contract document often dictates the dispute resolution process to be followed should negotiation fail. Normally this route includes one or more of the following:
1. Mediation, where an independent mediator is appointed to get the parties together to discuss and resolve the problem. Mediation is a non-binding process. The mediator is a facilitator to guide the parties towards an equitable resolution.
2. Arbitration, where an independent arbitrator is appointed to hear evidence and then make a ruling on what they believe is the correct solution. Arbitration is usually not binding, but this will vary depending on the Contract Document and the law of the country where the Contract is administered. Arbitration can be requested by either party. Demand for arbitration must be done within specified time periods. Once the employer has ruled on a Claim normally the contractor has to request arbitration within 30 days of the ruling they are unhappy with.
3. Litigation tends to take a more legal approach with the process focusing on the legal rights of a party, without necessarily understanding the project and the impact of Variations on the construction processes and Schedule.

Arbitration and litigation can be drawn out processes, often taking several years to resolve. They are also costly, and may not provide the answer either party was expecting. Legal disputes also often make the media headlines which could be damaging to the contractor's reputation and harm future work prospects.

Some projects require a Dispute Resolution Board (DAB) to be appointed at the start of the project. The DAB is usually made up of three members (one chosen by the contractor, another by the employer, and the third by the first two members). Both the contractor and the client are usually responsible for the costs of the DAB. The members of the DAB usually visit the project monthly and become aware of issues and problems as they arise. Should a dispute arise it's passed to the DAB for a ruling.

Depending on the country there may be other options available to contractors, so it's worthwhile consulting experts.

Unfortunately disputes which cannot be resolved do arise, and then it's important to follow the dispute resolution process stipulated in the Contract. Only as a last resort should contractors proceed down the legal route. Having said this, though, contractors should not hesitate to ask for a legal opinion or for expert advice. Of course also ensure that senior management is aware of the problems and the next steps being considered.

Just because there is a dispute doesn't mean the contractor can walk off the project. Sometimes construction companies do this, but it could be a fatal mistake if the proper Termination procedures haven't been followed, and may allow the employer to take action against the contractor for Breach of Contract.

Understand the laws in the state or country

Laws vary between countries and even between different states within the same country. It's important to understand the laws under which the Contract will be administered. Just because the project is located in a particular country doesn't always mean the laws of that country will apply. The Contract normally specifies the legal system that will be used in administering the Contract.

This could mean that if a dispute arises the contractor may have to argue the case in a foreign country, referring to a legal system they and their normal legal team aren't familiar with. This also results in additional travel costs to argue the Claim.

Before pricing a project the contractor should ensure they understand the legal basis of the Contract and where disputes will be heard.

Revising the Construction Schedule

After an Extension of Time Claim or Acceleration Claim has been approved the contractor should submit a revised Construction Schedule which shows the new approved Completion Date. This Schedule should not be changed in any respect from the existing approved Construction Schedule (links between tasks, task durations, or sequencing of the work) except for the changes agreed between the parties in the Extension of Time Claim or Acceleration Claim. In the case of Acceleration Claims these changes should reflect the new agreed construction methods, additional work days (when the contractor works on previous non-work days), revised sequencing of tasks and shorter task durations where relevant. This new Schedule must be approved by the employer in writing and it will become the Construction Schedule that progress and further Extension of Time Claims will be assessed against. This Schedule must have a unique revision number so that it can be referenced.

A revised Information Schedule which conforms to the new Construction Schedule should be issued to the employer.

Summary

1. Once a Claim has been submitted the contractor should ensure that it is dealt with speedily.
2. Claims should be tracked on a Claims Register which includes the date of the Claim, the reason for the Claim, the impact of the Claim and the Claim's status.
3. The contractor will inevitably have to respond to questions the employer raises in relation to the Claim. These need to be answered in writing.
4. In some circumstances it could be advisable for the contractor to make compromises to avoid the Claim becoming a dispute.
5. Unfortunately sometimes employers don't respond to Claims, or their Agents aren't impartial. Contractors need to manage these situations and take actions which are within the terms of the Contract.
6. Where disputes are unavoidable the contractor must follow the dispute resolution process in the Contract.
7. Before lodging a dispute the contractor must carefully consider the costs and other impacts of escalating the Claim to become a dispute and ensure they have exhausted all possibilities for resolving the Claim.
8. The contractor shouldn't hesitate in engaging expert advice.

Chapter 5 – Avoiding Claims and Disputes

Many contractors automatically assume that Variation Claims will be good for them - that it's easy money. They assume that because the Claim is legitimate that they'll be paid all their costs plus more. Unfortunately Claims aren't always the 'gold mine' or 'cash cow' that contractors think they will be.

Why Variation Claims aren't always good for contractors

Unfortunately Variation Claims aren't as profitable as contractors think they'll be because:
1. Contractors don't always receive the full value, or the time, they claimed in the Variation.
2. Often contractors forget, or aren't aware of some of the costs they incurred because of the Claim and therefore only claim and recover part of their actual costs.
3. Settling Claims can be a lengthy process taking several months or even years to resolve. Employers won't pay the value of the Claim until it has been settled and agreed. In the meantime the contractor has completed the work and incurred the costs. Some Claims can be for hundreds of thousands, or even millions, of dollars so consequently unpaid Claims can severely impact a contractor's cash-flow. Some contractors have even gone bankrupt waiting for payment for their Variation Claims.
4. Claims can be expensive to settle especially when experts and legal counsel have to be hired and often these expenses aren't recovered.
5. Claims distract management from managing the project which invariably costs the project elsewhere. They may even take management's time away from other projects, particularly when the Claim drags on long after the project has been completed and the team has moved.
6. An Extension of Time Claim results in the contractor's team remaining on the project longer. Often there isn't a corresponding increase in the value

of work, or maybe only a small increase. This affectively means the contractor's team is doing a smaller monthly turnover than was originally envisaged. In essence they'll probably be earning the contractor a lesser profit per month. Sometimes the contractor might not have another project to go to so they're happy to have their team paid to remain longer on the project. However, usually there are other projects that the contractor's team (people and equipment) could be moving to. Ultimately the Extension of Time means the contractor loses business opportunities and profit elsewhere.

7. There's always the risk that the quantum of claims may sour the relationship between the contractor and the employer. This may impact progress and morale on the project and also jeopardise future work opportunities with the employer.

8. An Extension of Time means that the completion date is achieved later which delays the release of retention monies and bonds. Since contractors often have a limited value of bonds they are able to raise they could be constrained in taking on additional new projects until the project is completed and the bonds and sureties are released.

9. Delays can be demotivating for the contractor's team. Supervisors and workers generally want to see progress on the project. A busy worker is usually a happy worker. Delays may mean that workers have time on their hands which often leads to them chatting and finding fault where none should exist.

10. Delays can be particularly troublesome on remote projects and projects where employees are living away from home. At the start of the project workers are usually told the project will be a particular length of time and they mentally prepare themselves to be away from their family for that time. When the period gets extended they become frustrated. Their families become upset, which puts pressure on workers and it can impact their productivity. Often these projects have a high employee turnover which can disrupt production as well as adding to the contractor's costs because of the additional mobilisation expenses for the replacements.

11. I've found that when a project suffers delays it takes the pressure off the project team on other non-critical areas and sometimes they relax which could negatively impact productivity and increase the contractor's costs. Costs that the contractor can't claim. It's therefore important to ensure that the team doesn't relax on the project and maintains maximum progress and productivity in areas not impacted by the delay.

12. On several occasions I've had projects that have started off slowly because of delays with access or late information. People mobilised have been unable to work at their optimal productivity as we struggled to find tasks to keep them busy. Unfortunately these teams became use to the relaxed pace of the project and when the information and access was finally received we battled to get them to increase their tempo of work. Invariably even once there were no longer impediments to the work, production was slow, costing us time and money.
13. When a Claim escalates to a dispute it often makes the media headlines, creating bad publicity for the contractor which harms their reputation and could damage their prospects of future work with other employers.

Ask these important questions

When the value of the project increase, or an Extension of Time is granted its important contractors make the following assessments:
1. Does the employer have the budget to pay for the increased value of work completed?
2. Will the employer's project still be viable? Particularly if the employer is a developer depending on the sales of the facility. If the project becomes unviable the employer may become bankrupt part way through the project leaving the contractor's bills unpaid.
3. Will the employer's Payment Guarantee cover the increased project value?

    For example: on one project the employer issued us a Payment Guarantee for the original project price of $20 million. The value of the project increased and the employer went bankrupt before the project was completed leaving unpaid bills. When we tried to claim money back from the bank we found they wouldn't pay any value in excess of the original contract value. Because of Variations we had completed $22 million of work. The employer paid us $21 million before going bankrupt. The bank stated the Payment Guarantee was for $20 million and since the employer had already paid more than this we weren't entitled to claim the outstanding $1 million from the Guarantee.

4. Has the project's insurers been notified that the value of the project has increased or that the project completion date has been extended?
5. Many guarantees and warrantees have a specified end date and these may need to be extended if the project Completion Date is extended.

Chapter 5 – Avoiding Claims and Disputes | 119

6. Are suppliers and subcontractors aware of the changes and can they accommodate them.
7. If the amount of work increases does the contractor have the additional resources to undertake the extra work?
8. Is the contractor able to sustain the revised cash flow? An increase in the value may mean that the contractor has to do a greater turn-over every month which might require additional funding. An Extension of Time usually means that retention monies are released later.

Is there an alternative?

When a change or delay occurs contractors are usually quick to submit a Variation Claim. But, occasionally there may be other options which could avoid the additional expense or reduce the delay. It's sometimes worth considering other options and discussing these with the employer. Obviously the contractor must be certain to ensure that they aren't taking responsibility for these options, but that they are merely suggesting alternatives to alleviate the situation. The contractor shouldn't place themselves in a position whereby they may be responsible for a product or design should they go wrong.

These alternatives could include:
1. Alternative construction methods.
2. Using different materials.
3. Changing the construction sequence.
4. Providing Partial Handover of sections of work to enable the employer to receive access on time.
5. Altering the design.

Avoid certain employers

Unfortunately some employers, Project Managers and construction companies have a reputation for being excessively contractual, being unreasonable or being overly enthusiastic with their claims – it may be sensible for the contractor to avoid working with these companies or people, or if they have to, then they should ensure their team is competent and able to defend the company.

Understand the employer's budget. If the project value is close to the employer's budget, or exceeds the amount, the employer will fight every Variation Claim even when it is legitimate. They may even lodge Claims against the contractor in an attempt to claw back money so the final project value is within their budget.

Some employers are disorganised, maybe not sure of what they want. In these cases the project could be delayed or incur numerous changes and Variations.

In some cases the employer's team (Engineers, Architects and Project Managers) may be inexperienced, disorganised or not very good. In this case it's almost certain the contractor will face delays and will be continually chasing missing information or dealing with faulty information – inevitably leading to Variation Claims. Trying to assist the design team and following up information is time consuming and can distract the contractor's team from managing the construction work which can ultimately impact progress and productivity thus harming the contractor.

Avoid particular Contract Documents

Some Contract Documents should be avoided because:
1. They are one sided and biased towards the employer, offering the contractor little protection should the project go wrong, the employer fail to pay for the contractor's work, when Variations occur or should there be a contractual dispute.
2. They contain ambiguous language where clauses are open to interpretation.
3. They are incomplete.
4. They don't include what was in the Request to Price or Tender Documents, or don't reflect what the contractor priced.
5. They have conflicting clauses which could lead to disputes.
6. They may result in one of the parties breaking the law.
7. They contain exculpatory clauses – clauses where one party (normally the party that drafted the Contract) is absolved of all blame or liability arising from the performance of the Contract.
8. The Contract is going to be administered in a foreign country whose legal systems the contractor isn't familiar with or where the legal processes may favour the employer. Contractors should remember it will be costly to travel to a foreign country and hire lawyers in that country to defend Claims which arise on the project.
9. They place excessive risks on the contractor – risks which the contractor can't control and risks which if they should eventuate could be catastrophic for the contractor.

Contract Documents need to be clear on the following:
1. Who the contracting entities are.
2. The Scope of Works.

Chapter 5 – Avoiding Claims and Disputes | 121

3. The parties' obligations under the Contract.
4. The price of the work.
5. The payment terms.
6. The Variation process.
7. The dispute resolution procedure.

Understand the tender or pricing documentation

Many contractors get into problems when they price the project because they haven't understood the Tender Documents or they misunderstood the Scope of Works and what their obligations are under the Contract. They commit to projects that they don't have the skills, experience or financial capacity to undertake. They accept Construction Schedules which are unreasonable or ones they don't have the resources to meet.

Errors in pricing invariably lead to problems constructing the project. Either in desperation to claw back their losses, or avoid Liquidated Damages, the contractor submits Variation Claims at every opportunity. Sometimes because they don't understand the Contract Documents they even misguidedly believe they are correct to submit some of these Claims.

Contractors must ensure they understand the pricing documentation, and that they prepare a thorough and accurate price based on the information provided and the information available to them after viewing the project site. Failure to do so will inevitably lead to mistakes and the potential for Variation Claims and disputes.

If anything is unclear in the pricing documentation queries should be submitted in writing to the client. If needed, the contractor should clarify any assumptions they've made as well as what their price is based on and what has been included and allowed for in their price.

If the contractor has to make assumptions and is unable to price the project properly with the information supplied it's probably best not to price the project.

Visit the project site before submitting the price (tender)

Some problems on construction projects can be avoided if the contractor visits the project site when they price the project. The purpose of these visits is to check:
1. For any restrictions or obstructions that could impact construction.
2. The ground conditions which are important if excavations are required. Rock can cause significant delays.
3. The access to the project site.

4. Whether the employer's activities, or the activities of their other contractors could impact construction.
5. If there's any ground water present that could impact excavations.
6. If there are restrictions to get vehicles or people onto the project site.
7. How existing structures could impact construction.
8. If existing utility lines (power, gas, water) could influence construction.
9. The location of service or utility connections that the contractor will require for their construction work.
10. The existing security of the site, including any requirements that the contractor will have to deal with during construction and any additional security measures the contractor will have to put in place to protect the works during construction.

Uncovering potential problems when pricing the project will allow the contractor to allow for them in their price and Schedule. Obvious problems which the contractor should have detected prior to submitting their price won't be cause for a Variation Claim and will be rejected by the employer.

Forewarning the employer of potential problems during the bidding or pricing process may allow them to adjust their documentation, budget and Schedule, preventing Variation Claims arising during construction.

The Construction Schedule

Contractors shouldn't rely on Construction Schedules provided by the employer, but should carefully scrutinise them to ensure that the logic is correct and that the time allowed for each task is sufficient.

The Construction Schedule should:
1. Be approved in writing.
2. Be in sufficient detail so that progress can be monitored and delays can be correctly recorded.
3. Have tasks correctly sequenced and linked so that if an item is delayed and this delay impacts other tasks that this impact can be clearly demonstrated.
4. Have logic which is clear to the employer. Where there can be doubt as to why the contractor chose a particular sequence or logic option (for instance to smooth out resource utilisation or to make use of available resources) this should be noted in the Schedule submission.

Chapter 5 – Avoiding Claims and Disputes | 123

5. Take into account all the known project and employer imposed constraints.
6. Allow for the normal average expected weather conditions unless the employer has accepted all weather delays.
7. Clearly indicate when the employer must provide access, permits, approvals, drawings and other information required from the employer to enable construction work to proceed and employer provided materials and items.
8. Allow sufficient time for commissioning.
9. Take into account the types, skills and quantity of the available resources.
10. Allow sufficient lead-time for the procurement of materials.
11. Be a reasonable assessment of the time required to complete the individual tasks, as well as the complete project, allowing for the safe completion of the project and permitting the delivery of a project with the required quality standards.
12. Accurately reflect the chosen construction methods and procedures.
13. Allow sufficient time for the employer to approve the contractor's designs and drawings (where applicable), and this approval time must be clearly articulated to the employer.
14. Where possible include the resources required for each task.

Unfortunately many disputes arise because of poorly constructed Schedules, or Schedules which don't accurately reflect the project conditions. Without a correctly formulated Schedule it is often difficult to substantiate Extension of Time Claims or Acceleration Claims.

Team work

My most successful projects were those where we as the contractor worked together with our subcontractors, the employer and their team, all as one team, focussed on delivering the project successfully. These projects were delivered safely, on time, within budget and with the required quality, and we made a good profit. Unfortunately some projects aren't managed in the same way and there's animosity between the different players from the start. There's a mentality of 'us and them – and not we'. It takes teamwork, a spirit of trust and cooperation and mutual respect to deliver a project successfully. Contractors shouldn't be out there to set traps, or attempt to trip employers and their team up. I've seen some contractors try and be clever and claim money they weren't entitled to, while they've forgotten to claim monies that were due to them. Employers need to treat their contractors fairly, paying them on time and for work they've done, not viewing all contractors as being crooked only out to get the employer's money. Both sides need to admit to their mistakes and keep egos away. Time spent arguing and fighting with other parties is usually time that could have yielded better results elsewhere on the project – time that could have prevented further problems from occurring on the project.

Unfortunately the way many construction projects are bid – awarding the project to the contractor with the lowest price – is a recipe for conflict. The low price is frequently due to contractors making errors when formulating their price, meaning they'll lose money and want to claw the losses back from the employer by some means. Alternatively the low price is a desperate ploy to win the project, hoping to then make money through Claims and Variations.

It's important that the project comes first, but also not at the expense of one of the parties. There shouldn't be losers on a project.

Communication

Communicating with the employer is essential. It's important to immediately notify employers when their decisions, or lack of decisions, are going to result in additional costs or delays. Often employers will specify something not knowing the implications, and then when they are informed of the consequences they could rescind the decision. Many times I've had employers surprised that their actions caused project delays and budget blow-outs.

No one like surprises, especially unpleasant surprises, and our employers are the same. Keeping them constantly informed helps prevent nasty and unexpected surprises. But simply talking to employers and subcontractors often gives contractors an insight to difficulties they are experiencing, or concerns they may have. Having advance warning contractors may be able to make better decisions so neither employers nor subcontractors are surprised by the contractor's actions.

The proper written communication of delays and Variations will form strong supporting documentation to back the contractor's Variation Claims.

Communication needs to be honest. Too often the employer's team doesn't want to tell the employer the truth – truth that the Schedule isn't achievable, that their budget will be exceeded or that their decisions (even sometimes their lack of decisions) will delay the project or add costs to the project.

Contractors also try and make the Construction Schedule fit the employer's Schedule. Sometimes contractors have to tell the employer that what they are asking cannot be done. Maybe offer alternatives or make suggestions on how the project could be modified to fit the client's requests. But just blindly saying 'yes' to every one of the employer's requests because the contractor doesn't want to offend the employer will inevitably lead to disappointment later, often resulting in Variation Claims.

Communication needs to be clear. Many problems on construction projects are as a result of misunderstandings or miscommunications. There should be no doubt as to what is meant. All too often emails and letters are sent in haste without being checked to ensure that the intent has been properly and accurately conveyed. Even misspellings can give a completely different interpretation to a letter.

Communication needs to be directed to the correct person using the correct address. Failure to do so may result in the communication becoming lost or not actioned.

Understanding the employer

Sometimes Claims and disputes can be minimised and even avoided when the contractor understands the employer and their needs. This understanding comes from talking to the employer as well as undertaking background research. Unfortunately people aren't always transparent and there are often hidden agendas – sometimes even personal agendas and egos. Knowledge of these agendas enables contractors to make more informed decisions which could avoid disputes. Knowing what is important to the employer may enable the contractor to reorganise the Construction Schedule to ensure the areas important to the employer are delivered on time. Providing the employer with choices and making them part of the decision making process could empower them, making them part of the resolution process.

Verbal communication

Verbal communication causes many problems on construction projects. Contractors could misunderstand or misinterpret these communications which then leads to conflict and disputes. There is no record of the communication so it could result in "I said, they said" situations. People can become very forgetful when things go wrong. Also, If there's no written record the contractor may forget to claim for a Variation issued verbally, or possibly even forget to carry out the Instruction causing the employer to be unhappy.

Some employers are notoriously coy about putting things in writing. Contractors should insist on Instructions being issued in writing. If employers refuse to issue a written Instruction which may have a time or cost impact then the contractor should confirm the Instruction, or request, in writing, submitting it to the employer so there's a record.

Contractors should ensure that Instructions discussed at Project Meetings are confirmed in the meeting minutes.

The contractor should make certain that their team is aware that they shouldn't accept any verbal Instructions from the employer or their team. In these instances they should politely ask the person issuing the Instruction to please issue the Instruction in writing to their manager. Unfortunately I have found on some projects that the employer approached our Supervisors, or even workers, and asked them verbally to make changes. Unless the contractor's management are aware of these Instructions they may not be recorded and the contractor will end up executing changes which they don't claim for.

The contractor's team – the contractor's early warning

The contractor's team must be aware of how critical it is to communicate with their Project Manager whenever there are events which could give rise to Extension of Time or Variation Claims. The contractor's Engineers and Supervisors are in the front line and are often the first to see a new or revised drawing. They are sometimes more aware how a small change has a major impact on their section of work. They'll instantaneously feel the impact when an area is handed to them that doesn't conform to the specifications, or when their section of work is negatively impacted by the employer's quality and testing regime which goes beyond what's considered normal. The contractor's team must report any potential delay or Variation to their management as soon as it's noticed.

Training

There's no better way of avoiding disputes than ensuring that the contractor's team understands the Contract Document. In the past when a project has involved a new form of Contract Document I've called in an expert to present a workshop on the Contract. In some cases it's also helpful to invite the team who will be administering the Contract on behalf of the employer to also attend the workshops. Having all parties familiar with the Contract Document can avoid misunderstandings and disagreements during the project – particularly when it comes to Variation Claims.

Should there be concerns that subcontractors won't be familiar with the Contract it's useful to invite them to a workshop to discuss the administration of their Contract.

Chapter 5 – Avoiding Claims and Disputes | 127

In any case it's important that the contractor's teams, including the estimators, attend workshops and lectures relating to contractual issues, the different forms of Contract and the submission of Claims. Having the correct knowledge will enable the team to avoid the many pitfalls encountered when pricing projects and managing them.

Subcontractor pricing

Many disputes arise between subcontractors and the General Contractor (Main Contractor). Some of these disputes are caused by poor Contract Documents and incomplete tender information. These disputes could result in delays to the project and additional costs for the contractor.

Often these problems start in the tender or pricing stage. Subcontract tender documentation will vary depending on the type and size of the subcontract. But, each should be as clear and complete as possible so there is no cause for misunderstandings which could result in quality, schedule and safety problems, or lead to Claims and Variations. The tender documentation should:

1. Have a clear Scope of Works.
2. Detail all the drawings and specifications that apply to the Contract (including a drawing register with all drawing numbers and revisions).
3. Contain the terms of the Contract, which would include payment terms and conditions, and insurance conditions.
4. Note special project conditions, including specific project labour agreements and wage rates.
5. Have specific requirements the contractor may have, such as particular staffing or equipment requirements.
6. Detail the project's safety, quality and environmental requirements.
7. Clearly state what the subcontractor will be supplying and what they will be provided with, this includes amongst other things, services (like power and water), storage and office facilities, cranes, scaffolding, off-loading facilities, security, and insurances, with any charges for using them specified.
8. Include the Construction Schedule, which highlights the subcontract activities and any discontinuities they should expect in the course of their work.
9. Note what spares and warranties are required and when these warranties should begin.
10. Include when commissioning should be completed and what tie-ins the subcontractor should allow for with other infrastructure.
11 Indicate all the documentation required from the subcontractor.

Often the contractor merely packages part of the Contract Document they have with their employer and passes this to their subcontractors with the pricing documentation. This may create confusion resulting in the subcontractor pricing items the contractor didn't intend them to price, or even misunderstanding the Document.

Selecting the subcontractor

The selection process for subcontractors is critical to the project's success. Subcontractors shouldn't be appointed solely on the fact that they are the cheapest, their price was used at tender stage, or that they're convenient to use. In selecting the subcontractor consider amongst many things the following:
1. Do they have experience with the particular work required?
2. Have they worked on similar projects or for similar employers? This not only relates to the type of work that is involved, but also to the type of project and employer. For instance, many subcontractors may be able to deliver a similar job on a commercial building project in the city, but have no experience on working on remote mining or oil and gas projects which have specific requirements and require more onerous safety standards.
3. Can they produce acceptable quality?
4. Do they have the financial means to carry out a project of this magnitude? Are they financially secure and not likely to face bankruptcy in the course of the project? Do they have Claims or disputes lodged against them which could impact their financial standing should they lose these Claims?
5. Have they undertaken projects with a comparable size in the past?
6. What is their track record on delivering similar projects?
7. What other work are they currently doing?
8. What is their safety record on other projects?
9. Do they have suitable resources available for the project?
10. Will they be able to honour warranties and guarantees?
11. Do they have a record of being Claims orientated or have they been involved in legal actions or disputes with other contractors and employers. Obviously this shouldn't be reason alone not to employ the subcontractor, but further investigation should be undertaken. The contractor should ascertain why there was a dispute or Claim. It's possible that these were legitimate.

Past performance, however, is not always indicative of how a contractor will perform on a project and I have, on occasion, had good subcontractors that have performed poorly, due to them being overcommitted on other projects, which meant they had insufficient and poor quality resources for my project.

The more that can be discovered about the subcontractor at this stage, the better. Not only can the capabilities of the subcontractor be researched, but it may be possible to discover their strengths and weaknesses which will, in turn, enable them to be better managed on the project. For instance, if a previous employer says the subcontractor submits lots of Claims, the contractor can implement steps to manage the Contract so as to ensure the subcontractor isn't given a reason to submit Claims. Research and talking to employers may enable contractors to discover who the subcontractor's best teams are, thus enabling them to request the subcontractor allocates these to the project.

Subcontractor documentation

The subcontract Contract Document is a legally binding Contract between the subcontractor and contractor with enforceable provisions on both parties. If the Contract Document is poorly worded, inconsistent or incomplete, it can lead to complications with the management of the subcontractor leading to Claims.

The subcontract Contract Document must include or make provision for:
1. The Scope of Works.
2. The price.
3. The Contract terms and conditions including; payment terms, invoicing instructions and retainage monies if required (including conditions for their release).
4. Specifications.
5. Applicable drawings.
6. Specific site conditions.
7. A list of deliverables required from the subcontractor.
8. The subcontractor's Construction Schedule which clearly shows commissioning and any discontinuities, and which complies with the overall Construction Schedule including meeting access requirements and allowing reasonable times for design and drawing approvals.
9. The project safety requirements.
10. Applicable quality requirements, procedures, tests and documentation.
11. Commissioning requirements.
12. Spare parts that the subcontractor must price.
13. Warranties and guarantees required, including their start dates.
14. Clarification of what the subcontractor must supply to carry out the works, as well as the contractor's obligations.

Much of the Contract Documentation should have formed part of the tender documentation.

Managing subcontractors

Some important points to note when managing subcontractors are to ensure:
1. The contractor's person managing the subcontractor understands:
    a. The subcontractor's Scope of Work.
    b. Both the subcontractor's and the contractor's obligations.
    c. How the subcontractor is reimbursed.
2. The subcontractor:
    a. Produces work of acceptable quality and are notified immediately when their work doesn't conform. Any non-conformances are recorded and defective work is repaired in accordance with the project specifications.
    b. Works according to the Construction Schedule.
3. The subcontractor receives access and information on, or ahead of schedule and isn't delayed by the contractor or other subcontractors.
4. Regular meetings are held with the subcontractor to discuss safety, quality and environmental matters, as well as progress on the project and any delays and Claims. The minutes of these meetings must be distributed to the relevant parties.
5. Subcontractors sign acknowledgement for the receipt of the drawings and information issued to them.
6. Where relevant, the subcontractor supplies shop drawings in accordance with the Construction Schedule, including allowing for obtaining the required approvals from the contractor or the contractor's employer.
7. Communication with the subcontractor of a contractual nature is in writing (any verbal instructions should be followed up in writing).
8. Only the contractor's delegated responsible employees communicates with the subcontractor.
9. Action is taken as soon as it appears that the subcontractor could be in trouble.
10. The subcontractor is forewarned of the contractor's intention to back-charge them for work or services supplied by the contractor and that these charges are invoiced regularly.
11. The subcontractor is paid in accordance with the Contract.
12. All guarantees and warranties are in place before the final payment is released.
13. Correspondence from the subcontractor is dealt with promptly.
14. Subcontractors don't begin work until there's a signed Contract in place and that they've supplied the required sureties and insurances.

15. The subcontractor's staff, equipment and their own subcontractors are approved by the contractor.
16. The subcontractor does not sublet portions of their work without permission.

Managing Design and Construct Projects

As discussed in Chapter 1 the contractor must closely manage the design process when they are responsible for the design to ensure that the design and the construction drawings are delivered in accordance to the Construction Schedule and that the design is in accordance with the employer's requirements as well as the applicable legislation and permit conditions.

If the design is prepared by an external party appointed by the contractor then the contractor must have a sound Contract in place with the designer. This Contract should include:

1. The Construction Schedule which clearly shows when information is required. This Schedule should allow the time required for the employer's approvals.
2. The Scope of Work including all the specifications, the employer's requirements and project constraints and design parameters.
3. Any particular construction methods or materials that the contractor may wish to use.
4. The responsibilities of the design team which could include:
    a. Preliminary investigations such as topographical and geological surveys.
    b. Liaising with the local authorities and obtaining permits and approvals.
    c. Ongoing design support.
    d. Quality inspections.
    e. Final certification of the project.
5. The risks and the responsibilities of the parties.
6. The insurances required.
7. Terms and conditions.
8. Design guaranties and warranties.

Regrettably many of these Contracts are poorly put together and there are often 'gaps' in the Contract where responsibilities aren't clear. These can cause delays as well as additional costs for the contractor. Sometimes because the Contract isn't clear the employer has cause to request that drawings and designs are modified because they don't meet their requirements. If these modifications are due to the contractor's fault the designer will claim the additional costs from the contractor and the contractor will be responsible for any delays to the project.

If the contractor manages the design process and monitors the design they should ensure that the design satisfies both the employer's requirements, while also being constructible, making efficient use of their available construction resources and enabling the facility to be constructed safely within the specified time period.

To avoid Claims against the employer the contractor must ensure that the employer approves drawings and the design in accordance with the Schedule. Also the employer must be timeously notified if they request drawings and designs to be modified where these modifications will cause a delay or Variation.

Shop drawings

Some items installed in the construction process require shop drawings before they can be fabricated. At times it's a Contract requirement for these to be submitted to the employer's representative for their approval, sometimes even in a particular format and layout. The contractor should ensure these requirements are included in the subcontract Tender and Contract Documents. I've, on occasion, had subcontractors submit drawings with the incorrect format which were rejected by the employer resulting in a delay.

Delays with submitting or approving shop drawings may cause delays to the project, so it's important that suitable and appropriate staff:
1. Monitor the production of shop drawings.
2. Check to ensure the item conforms to the dimensions, specifications and quality standards.
3. Submit the drawings to the employer for comment, if required.
4. Ensure the employer returns the drawings timeously.
5. Check that modifications or changes requested by the employer are within the Contract Scope. The employer will have to be notified if they have requested changes which don't conform to the Contract Document and which will result in a Variation Claim or a delay.
6. Return the drawings with appropriate comments to the fabricator or subcontractor.
7. Monitor the corrections to ensure that they are done timeously.
8. Check and approve the modified drawings.
9. Ensure there's a formal procedure to track the submittal and receipt of shop drawings.
10. Ensure drawings are submitted to the correct person, using the correct address.

Many contractors don't check the shop drawings since they believe it's the employer or their Designers responsibility. However the Designer often only goes through a 'rubber stamping process', and expects the contractor to have confirmed the item will conform to the project requirements. The contractor may be liable for mistakes on the shop drawings which lead to the fabricated item not meeting the project requirements.

> *Case study: On one of my projects the fabricator of a steel frame submitted a shop drawing of the item for approval. Our Engineer failed to check the drawing and simply gave his approval for the work to proceed. The dimensions of the item on the drawing though were smaller than required, so consequently when the item arrived it was too small, and didn't fit the structure. It had to be refabricated, resulting in additional costs to us and a delay in completion of the project.*

A copy of all shop drawings must be kept on site. When the relevant item is received it should be checked and compared with the drawings to verify it has been fabricated correctly.

Samples and mock-ups

Providing samples, or constructing mock-ups, allows the contractor to present a sample of the finished product to the employer for their approval. Often this is an opportunity to agree on the quality of the product as well as to resolve other issues. It's important these samples and mock-ups are accepted by the employer in writing and that they are kept until the project is completed.

On occasion employers have been known to ask for changes when presented with these samples and mock-ups. If these changes result in a product above the specifications or quality requirements included in the Contract Document, and where the contractor will incur additional costs or delays then the contractor should immediately give notice that if these changes are implemented there will be a Variation Claim. This notification will provide the employer an opportunity to rescind their requests.

By agreeing products and quality before work begins often means there won't be arguments later over the suitability and quality of the final facility.

The win-win approach

It's nice when a contractor is paid everything in their Variation Claim that they are entitled to. However, disagreements will arise that aren't completely clear or defined in the Contract Document, or that aren't sustained in the supporting documents. In these cases it's often necessary for a compromise to be reached between the parties, possibly an agreement where the contractor isn't awarded the full value of their Claim, but a solution where they don't lose.

Sometimes contractors have to view their Claim in relation to their overall Claims and the final result they would like from the project. From time to time it's prudent to make some concessions to appease the employer and the employer's team, while looking at the end game. The best result is one where no one feels they are losers. Losers can become vindictive which may negatively impact the outcomes of the contractor's other Claims. Even minor concessions can make the other party feel they've had a win. Having said this though, when contractors have a substantial legitimate Variation Claim they shouldn't be bullied into accepting a reduced Claim, below their costs, or even accepting a rejection of their Claim because they fear they'll upset the employer.

In some cases the employer may have already agreed and settled other significant Claims so the contractor may take a judicious view and withdraw small Claims which the employer feels strongly about. This will show a spirit of goodwill and shouldn't be seen as a sign of weakness.

So too, when the contractor deals with their subcontractors' Claims, these should to be dealt with fairly. Contractors shouldn't try and make money at their subcontractor's expense, nor should they necessarily pursue a good subcontractor, or a subcontractor they're depending on for every last cent. Contractors are often in a position of strength when it comes to paying their subcontractors. This position can give some contractors the feeling that they can disregard, refute or reduce subcontractors Claims, even when they're genuine Claims. Unfortunately these bullying tactics can backfire causing subcontractors to be unwilling to put extra effort into the project when required, or it could make these subcontractors reluctant to work with the contractor again, meaning they may not quote on future projects, or their prices will be increased to make allowance for the contractor's unreasonableness on previous projects. In some cases refusal to pay a subcontractor's genuine Claim, or delaying payment for a Claim while it's being disputed, could cause the subcontractor to become bankrupt with severe implications for the contractor and their project.

The solution to most problems will require some compromise. Often problems and disagreements arise which aren't clearly defined in the Contract. The goal shouldn't be to win every situation, but rather to solve each issue with minimal impact to all parties as well as to the project. The contractor's objective should be to resolve the Claims, problems and disagreements as speedily and as effectively as possible.

Take it higher if necessary

Often contractors are dealing with the employer's appointed representatives or Agent who are tasked with administering the project. These could be Architects, Engineers or Project Managers. On occasion these representatives may be incompetent. Sometimes they are working on a limited budget themselves so don't devote sufficient resources to the project. This poor management will result in delays and extras which will cost the employer more money. If all attempts to get the required information on time have failed it may be necessary to involve the Architects, Engineers or Project Manager's senior management. Often their senior management aren't aware of what's happening in the field. In fact we often find that their field representatives aren't giving their management the correct information and are usually blaming the contractor for delays and extra costs – the contractor is getting a bad name and being blamed for other people's incompetence.

If necessary the contractor should call a meeting with senior representatives of the employer's Engineers, Architects or Project Managers to express the contractor's frustration with the company's lack of performance. Their people in the field may be wounded with this action, but it's important to resolve these issues to rescue the project and for the contractor to maintain their good name, even if it means upsetting a few people.

If this doesn't help, talk to the employer. They often don't know the truth, and in many cases are being fed lies which blame everything on the contractor. Hopefully when the employer understands the problems they are savvy enough and competent to take action and get their Engineers, Architects or Project Managers to supply quality information on time so the project isn't delayed further.

Contractors shouldn't step back and let an incompetent professional team, or in some cases and inexperienced employer delay the project and cause additional costs. It's in nobody's interests and could negatively impact everyone's reputation. However, caution should be taken that the contractor doesn't make decisions that aren't in their Scope. Even if these are taken in good faith and in the project's best interests the contractor shouldn't be taking on responsibilities for which they could later be held liable for. Nor should they perform additional work without an Instruction because they might not be paid.

Bogus Claims

Some contractors see Variation Claims as a 'get-out-of-prison-card'. Often when contractors are behind Schedule, or are losing money, they think by submitting a Variation Claim it will solve their problems. Contractors will then often submit Claims based on improper or incomplete evidence and facts. These spurious Claims are usually quickly dismissed by the employer. However, they waste the contractor's time which could be better focussed elsewhere on the project. In addition the contractor's project team can be overly reliant on winning their Claim, pinning their hopes on a miraculous outcome, which then distracts them from finding other solutions. These bogus Claims are also often used as a foil to convince the contractor's senior management that the project doesn't have a problem and that things will turn out fine once the employer agrees to the contractor's Claim. This of course doesn't happen, so the contractor's senior management is caught off guard when the Claims are refuted and the project team reports the inevitable problems and losses. By then it's inevitably too late to save the project.

These Claims also irritate the employer. Spurious and bogus Claims create the impression that the contractor is dishonest. Genuine Claims submitted later by the contractor may be dismissed by the employer on the assumption that it is just as frivolous as the previous Claims. To be successful with Claims the contractor should approach their employers with honesty and integrity giving them little to doubt that their Claims are genuine.

Furthermore employers have been known to take future projects elsewhere because of their frustrations with the contractor's spurious Claims. This obviously has long-term repercussions for the contractor and their reputation.

The role of the contractor's management

Sometimes the contractor's senior management aren't aware of Claims that the project team are submitting. They may then be caught unawares by an irate employer who is frustrated by a particular Claim. Claims can also have a negative impact on the company's reputation. It's therefore important that the contractor's senior management are aware of what Claims the project team has submitted as well as Claims that they are considering submitting. With this knowledge they are often better able to face the employer, hopefully having sufficient information to be able to back the Claim and blunt the employer's attack, possibly even convincing the employer of the merits of the Claim.

If senior management are provided with a Claims register for each project, they will have a better understanding of what Claims have been lodged. Multiple Claims are indicative of a project that may be in trouble – either from the contractor's side or from the employer. If senior management become involved they could uncover the underlying causes of the Claims which can allow them to take action, such as engaging directly with the employer, so that the risk of further Claims can be lessened.

Tracking the progress of Claims to understand what Claims haven't been resolved also enables management to become involved in the resolution of the outstanding Claims – possibly engaging with the employer's senior management.

Unfortunately senior management are seldom aware of Claims that subcontractors have submitted. Subcontractor's Claims can be indicative of poor management of the project by the contractor's team. Failure to resolve subcontractors' Claims can also jeopardise the project's progress and lead to disputes. Often the additional costs of these Claims are only added to the contractor's cost reports when the Claims are resolved – which may result in unpleasant surprises for the contractor's management at the end of the project when the Claims are finally agreed in the subcontractor's favour adding to the costs. Claims which aren't promptly resolved may lead to legal action which can further add to the contractor's costs. It's therefore advisable that the contractor's senior management is provided with a register of subcontractor Claims.

Ask questions

Often disputes and Claims arise because of misunderstandings. If something isn't clear in the Contract Document, in the specifications or on construction drawings provided by the employer the contractor must ask the employer to clarify it. Often contractors assume something incorrectly, leading them to make mistakes that a simple question could have avoided. As discussed in Chapter 2 it is essential that these questions and the answers to the questions are in writing and recorded as an RFI, Engineering Query or similar. Simply because the information wasn't clear or was missing doesn't absolve the contractor from performing the work correctly and as the employer intended. Guessing information can be very expensive, sometimes requiring work to be redone with additional costs and possible delays. This often gives rise to Variation Claims. It's therefore essential that the contractor ensures that their team asks questions when they are uncertain of something and raise a query when there are discrepancies or an item doesn't appear to be correct. The contractor's Supervisors are often in the forefront when they work from drawings and specifications the employer has issued. In the rush to complete the work they sometimes misinterpret or misunderstand what is on the drawing, or, in some cases even make their own plan when something on the drawing is obviously incorrect.

It's also important to query discrepancies between drawings or between drawings and the specifications.

The contractor should not assume anything when pricing a project. If the Scope isn't clear, specifications aren't included, or there are contradictions between drawings, documents or specifications then it's important the contractor seeks clarity before submitting their price. Should the employer not provide the information requested then the contractor should include with their price the assumptions they made in arriving at their price, or ensure that it is clear what they have priced, what they are providing and what the employer should provide. In some instances the Contract Document and Scope of Works may be so poorly prepared that the contractor should consider not pricing the project. Being awarded a project where the pricing documentation is seriously flawed will almost certainly lead to a project with numerous Variation Claims and Disputes.

Take reasonable steps to prevent Claims from arising

Contractors can take reasonable steps to prevent Claims from occurring. These could include:

1. Issuing a Construction Schedule which clearly specifies when access is required from the employer.
2. Issuing an Information Schedule which is linked to the Construction Schedule and indicates when information is required from the employer. This schedule should be updated regularly and presented at project meetings and should preferably be included with the minutes of the meeting.
3. Warning the employer ahead of time when information or access is required. The contractor's aim shouldn't be to trap the employer – waiting for the employer to provide the information or access late so they can then lodge a Claim – but rather to be proactive to ensure there are no reasons for a Claim.
4. Checking the access granted immediately to ensure it meets the specifications and requirements.
5. Reviewing construction drawings and information as soon as they are received to verify the information is complete and sufficient to use for construction.
6. When possible cooperating with the employer to reduce the likely impact of the Claim.
7. Informing the employer as soon as a delay or Variation arises so that the employer can take action to mitigate, or minimise, the time and cost impacts.
8. Following up to ensure shop drawings and designs prepared by the contractor (or their subcontractors) are approved and returned as quickly as possible, and certainly within the specified maximum time allowed.
9. Notifying the employer of when the employer supplied materials and equipment will be required on the project, and then following-up to ensure these will be delivered on time.

10. Ensuring the employer meets their contractual obligations and taking immediate action should it appear the employer won't meet these obligations on time.
11. Regularly communicating with the employer and their team.

Give notice before the Contract Value is exceeded

Contractors should be updating their Claims and measured work as the project proceeds. The employer should be forewarned before the Contract Value is exceeded. Failure to do so could result in the contractor performing work which their employer doesn't have the money to pay for. This could result in the contractor not being paid at all, or as a minimum payment is delayed while the employer obtains the necessary funds. In some cases the completion of the project may be delayed while the additional funds are obtained.

Not being paid can have severe implications for the contractor's cash flow, while temporarily suspending the project while the contractor waits for payment will result in the contractor's resources standing idle, costing the contractor money.

Not being paid invariably leads to disputes and complications and is something contractors must be aware of and take active steps to avoid.

Don't exceed the value of Variation Orders

Variation Orders normally have a specified value. It's important the contractor avoids exceeding the value and timeously notifies their employer if it's likely the value will be exceeded. The contractor could very well not be paid the amount that exceeds the Variation Order. Even if they are paid, its possible payments will be delayed while another Variation Order is issued for the amount that exceeded the original Variation Order.

Variation Orders should be dealt with in terms of the original Contract and contractors should follow the same Variation Claim procedures should the work contained in the Variation Order change or be delayed.

International Contracts

Some Contracts are administered under the laws of a foreign country. These laws may be unfamiliar to the contractor, or may offer little protection to the contractor should a dispute arise. Furthermore, when a dispute is referred to the courts in that country, the contractor will have to travel to that country to argue their case. This will entail additional expense and time. Therefore where possible Contracts should refer to the laws of a country the contractor is familiar with, preferably their home country.

> For example: In a recent project in Western Australia the court cases have been argued in three different countries across two continents. This has entailed significant additional legal and travel costs for the parties resulting in some of the aggrieved parties having to settle for lesser amounts than they may have been entitled to.

In addition some of the judgements in the foreign countries may not be enforceable or legally binding in the project's country of origin.

Ask for help

It's important that contractors ask for help when they encounter a difficult, large or complex Claim, or a Claim that involves a subject they aren't familiar with. This help could be discussing the Claim with colleagues who have had similar Claims or have dealt with the employer and their team previously. Just discussing the Claim can help the contractor understand the problem better, possibly giving them a new angle to approach the Claim, or providing additional supporting documentation and facts.

If in doubt over how to proceed with a Claim contractors should consult experts to advise them. Every Contract is different so what worked and was correct on a previous project may not be the same on the current project. Asking for expert opinion will cost money, but it could very well avoid an expensive failed Claim.

Asking for help may also prevent the contractor from lodging Claims which are weak or cannot be supported.

Summary

1. Claims aren't as lucrative as contractors think they are. Often Claims are rejected, the contractor fails to claim all of their costs or they don't fully understand the full impact of the Variation. Claims take time to prepare and even more time to successfully negotiate a final settlement.
2. It's often better for the contractor to avoid certain projects, clients or design teams that could give rise to Variation Claims.
3. Contractors should be proactive to assist the employer so that Claims can be avoided or minimised.

Chapter 5 – Avoiding Claims and Disputes | 141

4. Team work and good communication is essential to minimising Claims. A sound knowledge of the Contract Document and contractual procedures is necessary.
5. The proper selection of subcontractors as well as appointing them using sound Contract Documents and then managing them properly will minimise the risks of the subcontractor lodging a Claim against the contractor.
6. Contractors should avoid submitting bogus Claims or Claims that have been excessively inflated as these can adversely impact their reputation and cast doubt on their legitimate Claims.
7. Ask questions when drawings, specifications or the Contract Document isn't clear or when there is conflicting information.
8. Understanding the Contract Documents and Scope of Work and ensuring the price submitted clearly states what is included in the contractor's price can assist with avoiding misunderstandings which could later form the basis of a Claim.
9. When dealing with large or complex Claims, Claims relating to an unfamiliar Contract Document, or an unusual topic, then the contractor should ask experts for assistance and advice.

Conclusion

Contractors need to be contractually astute. They must understand the terms and conditions of the Contract when they price the project, ensuring they are acceptable. These terms and conditions should be checked before the Contract Document is signed to ensure they are what the contractor priced. The contractor's Project Manager should understand the project's Contract Document and ensure they administer the project in terms of the conditions contained in the documents. By being proactive and knowledgeable most disputes can be avoided.

The project team should have knowledge of administering the Contract, and where necessary should attend specific training courses related to the Contract.

Almost every project will have changes. It is how these changes are dealt with that could determine the project's outcomes and success. Contractors have every right to Claim for legitimate delays and Variations. Failure to submit Claims will result in the contractor doing work at their cost. Furthermore, failure not to Claim for an Extension of Time that the contractor is entitled to could result in the contractor not completing the project by the contractual Completion Date which could allow the employer to claim Liquidated Damages from the contractor. In addition failing to complete the project on time without a legitimate and proven cause could result in the contractor's reputation being tarnished.

It's usually too late to submit Claims at the end of the project when the contractor realises that they are in trouble and that the employer isn't planning to help them out. At this late stage their Claim could be Time-Barred in terms of the Contract. In addition the contractor might no longer have access to all the information pertaining to the Claim, or the people who were directly involved with the Claim could have moved on to other projects or companies.

Claims that are clear and concise, and which have facts which are supported and can easily be justified and which are submitted within the time periods specified in the Contract Documents are often agreed and settled with employers with little effort. However, the Claim process begins long before the actual Claim is lodged – the success of most Claims will depend on the supporting documents including; correspondence, records, the agreed Construction Schedule, minutes of meetings and drawings. Without verifiable supporting documents the likely success of the Claim will be compromised.

Accurate project documentation assembled from the start of the project could enable the contractor to defend Claims that the employer or their subcontractors may lodge against them and will also form the basis of Claims the contractor raises.

From time to time the contractor may have reason to submit a Claim against their subcontractor. These Claims should be in terms of the subcontract Contract Document and the subcontractor should be timeously warned of these Claims and back-charges and they should be agreed as soon as possible. Also, subcontractors will submit Claims to the contractor. Some of these Claims may be for the contractor's employer and the contractor needs to ensure that the Claim is presented to their employer in compliance with their Contract and it is presented in a logical manner with supporting documents using the correct rates and includes the contractor's costs and mark-up related to the work. Claims from subcontractors which are directly attributable to the contractor's actions should be fairly and promptly dealt with.

Some Claims can be large and complicated and contractors should not hesitate to engage expert advice and help. This book serves as a guide only, as do most other books.

Once the Claim has been submitted to the employer the contractor may still have a lengthy process to negotiate and convince the employer of the legitimacy and the quantum of the Claim. Contractors should not be put off in this process by employers that try to bully them and threaten them that they won't be working on their future projects. However, contractors do need to listen to genuine concerns the employer may have. Questions need to be dealt with promptly ensuring the contractor presents the answers and supporting facts in a logical manner. The Claims settlement process must remain civil, steering clear of personal insults and feelings. It may be necessary at some stage to compromise, considering what has been gained and what is still outstanding – in particular considering the 'bigger picture'.

Contractors should not be too proud to admit when they are wrong. To continue fighting about something that is wrong is usually a waste of effort, irritates the other party and compromises the contractor's integrity.

Once contractors get a reputation of submitting inflated or bogus Claims it can become difficult to convince the employer when a reasonable and just Claim is submitted. Even legitimate Claims can be viewed as bogus if the contractor is unable to convince the employer they are legitimate and substantiate them with all the facts and documentation.

Conclusion

While contractors often consider that they make money out of submitting Claims, they often don't consider the cost and effort that goes into preparing and submitting the Claims – effort that could possibly have been used profitably elsewhere. In addition contractors can damage their reputation if they become known for submitting Claims. I would therefore prefer to avoid Claims as far as possible by ensuring we work for employers that are reasonable and fair and with strong design and project management teams. Where ever possible the contractor should work with the employer to avoid delays and Variations as far as possible.

However, if the contractor has legitimate reason in terms of the Contract to submit an Extension of Time Claim or a Variation Claim and they have all the supporting evidence they should do so.

Glossary

Terminologies vary between different construction industries, countries and even companies. The descriptions below relate more to their meaning within the book and aren't necessarily their official descriptions.

Acceleration – to shorten the Schedule, or programme, so the project is completed earlier, or alternatively, to complete more work in the same time period.

Access – contractors require access to carry out the work in their Scope. The work area must be provided to the contractor in the condition specified in the Contract Document.

Bid Documents (Tender Documents, Request for Price) – documentation the employer provides to the contractor so they can price the construction project. The Bid Documents could also be the documents the contractor provides to their subcontractor to price their section of work.

Bonds – a form of guarantee issued by a bank or insurance company to insure the client (up to a specified value) should the contractor fail to fulfil their obligations as detailed in the contract.

Breach – is when one of the contracting parties fails to fulfil their obligations in terms of the Contract. Breaches can be minor or major.

Cardinal Change – is a change that requires the contractor to perform work that's radically different to that which was envisaged in the Contract Document.

Change Order (Variation Order) – the written agreement between the parties which sets out the costs and Scope of additional work, or the change to the Contract.

Claim – a demand from one of the contracting parties for adjustment to the Contract.

Claim Register – a schedule or list of all the Claims

Completion Date – the date specified in the Contract and the approved Construction Schedule when the project must be completed.

Concurrent Delays – are two or more Delay events that overlap. These Delays may be caused by the employer, the contractor or events outside the control of either party.

Construction Drawings – the drawings issued to the contractor to enable them to construct the project.

Construction Schedule (construction program/programme) – the Schedule which the employer has agreed is the official one for the project and which sets out the sequence and duration of the construction activities. It's used to measure progress, adjudicate any Extension of Time Claims, and if necessary, to quantify the amount of the Liquidated Damages.

Constructive Acceleration – is when the contractor accelerates the work to make up the lost time due to a delay event outside their control, but the employer doesn't issue an instruction to Accelerate, nor accept the contractor's Extension of Time Claim, but insists the project is completed in accordance with the original completion date. The instruction to Accelerate is implied by the employer's insistence that the project is completed on time despite the delays or increased Scope.

Contract – is the agreement between two or more parties. In construction there is usually a Contract for the contractor to construct a facility and the employer to pay the contractor for the work. The Contract would outline the terms and conditions of the agreement as well as the obligations and rights of the contracting parties.

Contract Document – are all the documents that form the basis of the Contract between the employer and the contractor. The Contract Document must be signed and agreed by both parties.

Contract Value – is the value of the work that the contractor is employed, or contracted, to complete for the employer. The Contract Value can be varied (both up and down) by the addition of Variation Orders or Change Orders.

Contractor – a company that constructs or builds a facility or a portion of the facility for the employer.

Cost-Plus (cost-plus a fee) – when the contractor is reimbursed their actual costs incurred in carrying out the Contract, or Variation, as well as a mark-up on these costs which is proportional to the costs and is normally expressed as a percentage.

Critical Path – **is** a sequence of activities in the Construction Schedule which are linked, and whose delay will affect the overall project completion.

Daily Log (Daily Diary) – a daily record which is usually prepared by the contractor and records what occurred that day on the project.

Damages – are the costs or losses that one party suffers, often because of the actions of another party. These costs could be recovered in a Claim for Damages.

Design and Construction – construction projects where the contractor is employed to design and construct the facility.

Dispute Resolution – the process whereby a dispute is resolved.

Disputes – when a difference between the contractor and the employer, or other parties to a Contract, cannot be resolved amicably and it has to be referred to third parties for resolution. The Dispute could arise because the parties cannot agree that a Variation Claim submitted by one party has a legitimate basis, or they cannot agree the quantum of the Claim.

Disruption - is the loss of productivity caused by the disturbance, hindrance or interruption of the contractor's construction methods and their planned normal production, which results in lower efficiencies.

Drawing Register – a record of all the drawings (including superseded drawings) issued on the project and their date of issue.

Employer – the party who employed and contracted the contractor. The employer may be the owner of the facility, the managing contractor, or another contractor. Normally the employer is the party that pays the contractor.

Extension of Time – is the additional time required to complete the project because a delay, not caused by the contractor, impacted the project's Completion Date, or because additional Scope or work was added.

Float – the amount of time that a task can be delayed without impacting the final project Completion Date.

Force Majeure – unforeseeable course of events which none of the contracting parties has any ability to prevent.

Formwork (shutters) – the forms or structures used to shape and contain the wet concrete used in structures until it has gained sufficient strength to support itself.

Hold (1) – A term which is often inserted on a drawing by the designer to indicate the information for a particular section is incomplete, or isn't approved yet, so the contractor cannot use the information. Often the section on Hold is denoted by a cloud encircling the section on the drawing.

Hold (2) – The employer may put a section of work On Hold, which is an Instruction to the contractor to stop work. This should be issued in writing.

Information Required Schedule (Information Log or Information Register) – a schedule, or list, of when the client must make information available so as not to delay the contractor.

Instruction (Site Instruction) – is a contractual request made by the employer. Instructions often cause the project to be varied and may result in a Variation Claim.

Insurance – cover for potential losses.

Glossary

Lead Time – the amount of time taken for an item to be delivered to the project once it has been ordered. This time includes the time to design, manufacture and transport it to site.

Liquidated Damages – are a specified amount of money which the contractor will pay the client should the contractor fail to meet the agreed Contract Completion Dates.

Overheads (Overhead Costs, indirect costs) – overhead project costs are costs the contractor incurs to run the project which cannot be directly related to specific tasks. This usually includes the provision of management, supervision, site facilities, insurances and bonds. Company overheads are the costs a company incurs which are not directly attributable to a specific project, but are related to running the company and include costs such as Head Office rental, management and various support departments, such as finance and tendering.

Partial Access – in some cases the contractor and the employer may mutually agree that the employer takes access of part of the works, or access of incomplete work, so that they and their other contractors can proceed with their activities while the contractor continues to complete the facility. Granting Partial Access can be disruptive to the contractor's work, but it may allow the employer to take earlier access which could prevent the employer applying Liquidated Damages when the contractor is behind Schedule, or it could be a way of accelerating the project.

Preliminaries (Overheads) – this is the contractor's price for complying with the general obligations of the Contract. These costs cannot be allocated to a specific task or portion of the project, but are rather the costs associated with managing the project as a whole. These costs could include the project offices, management and supervision, insurance and bonds.

Project – any construction work.

Project Manager (site manager, construction manager or site agent) – is the person responsible to manage the contractor's work on the construction project.

Punch List (Snag List) – is a list of outstanding items or repairs that must be completed so that the facility complies with the employer's requirements.

Rates – cost. On some projects the contractor provides costs for various tasks, and these are included in the Contract Document.

Repudiation – is the refusal or failure to perform a duty or obligation in the Contract.

Request for Information (RFI) – the contractor's request for information, or a request to clarify information already supplied, so they can proceed with construction.

Glossary | 149

Schedule (often referred to as a programme, program, bar chart or Gantt chart) – a graphic representation of the timetable needed to complete the project, showing the sequencing, inter relationship and duration of the various project tasks and activities.

Scope of Work – the work which the contractor is contracted to do. The Scope normally takes the form of a written description of the work contained within the Contract Document.

Shop Drawings – drawings produced (normally by the contractor, their suppliers or subcontractors) to show the details of an item they have to fabricate.

Specifications – definitions of the materials, processes, quality, products and systems to be used in the works.

Standing Time Claim – is a claim for personnel or equipment that can't be utilised as planned and as per Schedule due to a delay or Variation.

Subcontractor – a contractor employed by a contractor to do a portion of their works.

Supervisor (foreman) – is the person who supervisors the contractor's workers, or a section of work.

Suspension – is a temporary halt to the Contract or a party's obligations in terms of the Contract. A contractor may wish to respond to actual or alleged Breaches of Contract by an employer by suspending works, or an employer may wish to respond by suspending payment. Suspension can be used by one party to allow it time to consider how to proceed with a project.

Tender (bid, estimate or quote) – a price, or quotation, to carry out work which is submitted by the contractor to the employer.

Tender Document (Request for Price (RFP), Request for Quotation) – are the documents issued to the contractor to enable them to price the project. These documents could include the Scope of Work, the terms and conditions of the Contract, specifications, drawings and ancillary project information.

Termination – is the dissolution of the Contract. Once a Contract is terminated it no longer exists.

Time-Bar – a clause in the Contract Document that sets a strict deadline by when a party may submit (and sometimes respond to) a Claim.

Variation Claim – a Claim to vary the Contract. The Claim could be for additional time and (or) for additional compensation.

Variation Order (Change Order) – the written agreement between the parties which sets out the costs and Scope of additional work, or the change to the Contract, or a change to the Completion Date.

Notes

References

Civitello, Jr. Andrew M & Levy, Sidney M. Construction Operations Manual of Policies and Procedures: 4th Edition, McGraw-Hill

Cushman, John; Carter, Robert; Gorman, Paul, Coppi, Douglas. Proving and Pricing Construction Claims. Aspen Publishers

Dykstra, Alison. Construction Project Management: A Complete Introduction, Kirshner Publishing Company, INC.

Glasov, Joshua. Liquidated Damages in Construction Contracts Part 2. Construction Law Today.

Gould, Nicholas. Making claims for time and money. Fenwick Elliot.

Haidar, Ali, D. Global Claims in Construction. Springer

Hall, Thomas, J; Smith, George Bundy. Interpreting Conflicting Contractual Provisions. The New York Law Journal.

Halpin, Daniel W & Senior, Bolivar A. Construction Management: 4th Edition, Hamilton Printing

Hewitt, Andy. Construction Claims and Responses: Effective Writing and Presentation. Wiley

Jackson, Barbara J. Construction Management Jumpstart: 2nd Edition, Sybex an Imprint of Wiley

Linares, Thierry. Time at Large and Extension of Time Principles. Expert's corner Paper 2013-01

Livengood, John. Concurrency World Tour. Navigant Construction Forum.

Livengood, John. Construction Claims from A to Z. Navigant Construction Forum

Mincks, William R & Johnston, Hal. Construction Jobsite Management: 2nd Edition, Thomson Delmar Learning

References

Mubarak, Saleh A. Construction Project Scheduling and Control. 2nd Edition: John Wiley and Son Schexnayder, Clifford J & Mayo, Richard E. Construction Management Fundamentals, McGraw-Hill

Nelson, Derek. What is Constructive About Acceleration. Paper published by Hill International

Netscher, Paul. Successful Construction Project Management: The Practical Guide. Panet Publications

Netscher, Paul. Building a Successful Construction Company: The Practical Guide. Panet Publications

Pinsent Masons. Termination and Suspension of Contracts. Out-Law.com

Pinsent Masons. Checklist for Terminating a Construction Contract. Out-Law.com

Richey, Jason, L; Wickard, William, D. Consequential Damages in Today's Construction Industry. Constructioneer, acppubs.com

Rubin, Robert; Fairweather, Virginia; Guy, Sammie. Construction Claims; Prevention and Resolution 3rd ed. Wiley

Shaw, Robert; Pentony, Tony. Proving Delay Claims (Walter Lilly & Co vs Mackay). Lavan Legal

Spinella, Frank. Conflicting Contract Clauses. NH Construction Law

The NSW Government Procurement System for Construction: Procurement Practice Guide. Handling Prolongation and Disruption Claims.

Walker, Anthony. Project Management in Construction: 5th Edition, Blackwell Publishing

Also by Paul Netscher
Successful Construction Project Management: The Practical Guide
Available from Amazon and other book stores

"This is a fantastic book to get a realistic and detailed idea of construction management."

Written by a construction professional, for construction professionals this invaluable book provides a step-by-step guide to successfully managing a construction project – including what-not-to-do to avoid costly mistakes. Learn how to master construction management and avoid the many pitfalls In construction. The Chapters include planning the project, starting it, scheduling, running the project, completing it, people, materials, equipment, quality, safety, subcontractors, contractual and financial.

Those who have read the book comment:
- 'I highly recommend this book be read by all newly qualified construction project managers as well as those more experienced.' (Reader on Amazon UK)
- 'It is quite clear beyond any doubt that Mr. Netscher has been through many projects of all different sizes, in different construction categories and in many different places which opens up to different cultural labor force and different construction method and added risk. I'm glad he didn't take the usual class room process group text book approach. Most chapters deals with a specific content, ready to be applied by the contractor. It's also good to learn from someone who made mistakes, takes blame, which made the book real. He presents the information to cover everything in an easy flowing read. Great Job. (Reader on Amazon)
- 'This is a fantastic book to get a realistic and detailed idea of construction management. It seems like it would be useful to people with experience, and it is very accessible to people like me that want to learn more about the field.' (Reader on Amazon)

This easy to read book is filled with practical everyday examples incorporating 28 years of construction experience gained on over 120 projects in 6 countries. **Many topics aren't included in other construction management books despite their importance.**

Ensure your next construction project is successful.

Also by Paul Netscher

Building a Successful Construction Company:
The Practical Guide

Available from Amazon and other book stores

"A practical guide based on the great experience of a real professional."

Many construction companies fail despite the hard work and knowledge of their owners. Is it due to bad luck?
This book is an invaluable guide for construction project managers and construction company owners. It explores essential solutions for creating a successful construction company and constructing profitable projects.

Those who have read the book comment:
- 'Highly recommended! Excellent read and packed full of valuable tips, as a 27 year construction person that is finally setting out on my own with a company I'm building from scratch, this book is a must read!!' (Reader on Amazon)
- 'A practical guide based on great experience of a real professional. If just only part of recommendations is followed by contractors the ratio of successful projects would grow dramatically. This book is recommended for owners reading to understand contractors backstage (helps establish proper negotiating strategy and reasonable tendering approach). Simple language provides understanding for construction business beginners. Case studies from real life and methodologies will be interesting for professionals.' (Reader on Amazon)
- 'Excellent writing, very useful and all around good read.' (Reader on Amazon)

The various chapters discuss the importance of picking the right job, finding work, tendering (bidding or quoting), winning the project, delivering the project, avoiding unnecessary costs, increasing revenue, financial and contractual controls, managing the company, the importance of good people, growing the company and ensuring the company has a good reputation. **Ensure your project and company's success using the invaluable tips and information in this book.**
A natural follow-on book to Successful Construction Project Management

Printed in Great Britain
by Amazon